BestMasters

Mit "**BestMasters**" zeichnet Springer die besten Masterarbeiten aus, die an renommierten Hochschulen in Deutschland, Österreich und der Schweiz entstanden sind. Die mit Höchstnote ausgezeichneten Arbeiten wurden durch Gutachterinnen und Gutachter zur Veröffentlichung empfohlen und behandeln aktuelle Themen aus unterschiedlichen Fachgebieten der Naturwissenschaften, Psychologie, Technik und Wirtschaftswissenschaften. Die Reihe wendet sich an Personen aus Praxis und Wissenschaft gleichermaßen und soll insbesondere auch dem wissenschaftlichen Nachwuchs Orientierung geben.

Springer awards "**BestMasters**" to the best master's theses which have been completed at renowned Universities in Germany, Austria, and Switzerland. The studies received highest marks and were recommended for publication by supervisors. They address current issues from various fields of research in natural sciences, psychology, technology, and economics. The series addresses practitioners as well as scientists and, in particular, offers guidance for early stage researchers.

Laura Wirth

Weighted Automata, Formal Power Series and Weighted Logic

 Springer Spektrum

Laura Wirth
University of Konstanz
Konstanz, Germany

ISSN 2625-3577   ISSN 2625-3615   (electronic)
BestMasters
ISBN 978-3-658-39322-9   ISBN 978-3-658-39323-6   (eBook)
https://doi.org/10.1007/978-3-658-39323-6

Responsible Editor: Marija Kojic
This Springer Spektrum imprint is published by the registered company Springer Fachmedien Wiesbaden GmbH, part of Springer Nature.
The registered company address is: Abraham-Lincoln-Str. 46, 65189 Wiesbaden, Germany

To my mom, for always being by my side,
and in loving memory of Opa Helmut, the biggest fan of my studies.

Acknowledgements

I would like to express my deepest gratitude to my supervisor Prof. Dr. Salma Kuhlmann for her patient guidance throughout my studies. Special thanks shall also go to my second supervisor Prof. Dr. Sven Kosub, whom I could always ask to clarify computer-scientific aspects of any kind. I had the advantage of being able to ask both of them for advice and answers to resolve difficulties at all times. Their remarks and comments have been a valuable help and powerful encouragement.

Furthermore, I would like to thank Prof. Dr. Manfred Droste for our exchange regarding Section 2.5. I could ask him to obtain references, and he gave me a hint that finally resulted in a proof of Theorem 2.5.19, which generalizes the classical result of Büchi, Elgot and Trakhtenbrot.

I owe a particular debt to Dr. Lothar Sebastian Krapp, who calmly supported me in overcoming all problems that arose during this work. His deep mathematical insight has given rise to many invaluable remarks and comments, which have essentially improved the presentation in many respects. In particular, I wish to thank him for taking the trouble to attentively proofread several iterations of this work. Beyond that, I am lucky enough to have him as a friend.

Most warmly, I thank my family – my sister Anna-Lena, my parents Bernd and Manuela, as well as Oma Anne and Oma Ria – for their less mathematical but more loving and emotional support, while I was studying and writing obscure theses that none of them is likely to ever read.

Finally, I wish to thank my fellow students Carl Eggen, Moritz Link, Patrick Michalski, including Philipp Huber, with whom I could share my enthusiasm for the subject of this work.

Abstract

A basic concept from Theoretical Computer Science for the specification of formal languages are finite automata. By equipping the states and transitions of these finite automata with weights, one obtains the quantitative model of weighted automata. The included weights may model e.g. the amount of resources needed for the execution of a transition, the involved costs, or the reliability of its successful execution. To obtain a uniform model, the underlying weight structure is usually modeled by an abstract semiring. The behavior of a weighted automaton is then represented by a formal power series. A formal power series is defined as a map assigning to each word over a given alphabet an element of the semiring, i.e. some weight associated with the respective word.

In this work, we put emphasis on the expressive power of weighted automata. More precisely, the main objective is to represent the behaviors of weighted automata by expressively equivalent formalisms. These formalisms include rational operations on formal power series, linear representations by means of matrices, and weighted monadic second-order logic.

To this end, we first exhibit the classical language-theoretic results of Kleene, Büchi, Elgot and Trakhtenbrot, which concentrate on the expressive power of finite automata. We further derive a generalized version of the Büchi–Elgot–Trakhtenbrot Theorem addressing formulas, which may have free variables, whereas the original statement concerns only sentences. Then we use the language-theoretic approaches and methods as starting point for our investigations with regard to formal power series. We establish Schützenberger's extension of Kleene's Theorem, referred to as Kleene–Schützenberger Theorem. Moreover, we introduce a weighted version of monadic second-order logic, which is due to Droste and Gastin, and analyze its expressive power. By means of this weighted logic, we derive an extension of the Büchi–Elgot–Trakhtenbrot Theorem. Thus, we point out relations among the different specification approaches for formal power series. Further, we relate the notions and results concerning formal power series to their respective counterparts in Language Theory.

Overall, our investigations shed light on the interplay between languages, classical as well as weighted automata, formal power series and monadic second-order logic. Hence, the topic of this work lies at the interface between Theoretical Computer Science, Algebra and Logic or, more generally, Model Theory.

Contents

1. Introduction

A fundamental concern of Computer Science is the study and processing of information
and data. Any information requires to be specified, and depending on the application
context, a suitable representation or specification is chosen in order to interpret the
information in a targeted manner. More precisely, the processing of data, and in
particular the amount of resources required for this, depends on the chosen method
or formal model for the representation of the conveyed information. Especially when
working with *infinite* objects or structures, it is essential to have a *finite* specification of
the conveyed information. In this work, we focus on the description-oriented research
in Theoretical Computer Science, rather than on algorithm-oriented results. However,
the two research areas really are highly, even inseparably, connected.

The information conveying structures that are considered in this work are formal languages and formal power series. Formal languages are sets of words over a given alphabet, and the formal power series, which concern us, are maps associating elements
of an abstract semiring to words over a given alphabet. Thus, languages model qualitative information concerning the membership of words, whereas formal power series
provide quantitative data about words. In fact, formal power series can be regarded
as quantitative extensions of languages.

In Theoretical Computer Science, the basic tool for the specification of languages
are finite automata. Historically, finite automata originate in the mid-1950s in the
work of Kleene [18]. In his fundamental result, which is commonly referred to as
Kleene's Theorem, Kleene characterized the languages that are recognizable by finite
automata as rational languages[1]. Mainly motivated by decidability questions, the
expressive equivalence of finite automata and monadic second-order logic was derived
independently by Büchi [4], Elgot [15] and Trakhtenbrot[2] [43] in the early 1960s. Their
equivalence result, referred to as Büchi–Elgot–Trakhtenbrot Theorem, establishes a
very early connection between the theory of finite automata and Mathematical Logic.
The two approaches often complement each other in a synergetic way, and their relation
is highly relevant for multiple application domains, e.g. in verification and knowledge
representation, for the design of combinatorial and sequential circuits, as well as in
natural language processing. In Theorem 2.5.19, we further present an extension of
the Büchi–Elgot–Trakhtenbrot Theorem to monadic second-order formulas, which may
have free variables, whereas the original result addresses only sentences. The proof of

---

[1] A language is called rational if it can be constructed from finitely many finite languages by applying
the rational operations union, concatenation and Kleene star.
[2] Due to different transliteration from the Cyrillic, various spellings of this name are common in Latin
script.

Theorem 2.5.19 is a new contribution of this work[3]. Overall, we are provided with three expressively equivalent tools for the specification of languages: finite automata, rational operations, and monadic second-order logic.

However, the main focus of this work concentrates on the study of formal power series. On the one hand, formal power series support the modeling of quantitative phenomena, e.g. the vagueness or uncertainty of a statement, length of time periods, or resource consumption, whereas languages are not suited to account such subtleties. On the other hand, formal power series constitute a powerful tool in relation to languages and automata, since they, in a sense, lead to the *arithmetization* of the theory. In addition, formal power series are of interest in various branches of Mathematics, particularly in Algebra. We therefore consider extensions of the above-mentioned language-theoretic specification tools to the realm of formal power series. By equipping the transitions and states of classical finite automata with weights, we arrive at the concept of weighted automata. The behavior of a weighted automaton is represented by a formal power series. Weighted automata provide a quantitative model in the sense that they take into account weights associated to computations. Therefore, weighted automata are employed for the description of quantitative properties in various areas such as image compression, probabilistic systems and speech-to-text processing. The notion of a weighted automaton was first introduced by Schützenberger [37] in 1961. Furthermore, Schützenberger was the first to investigate rational formal power series in the context of languages and automata. In particular, he proved an extension of Kleene's Theorem, referred to as Kleene–Schützenberger Theorem. The results of Kleene, Büchi, Elgot, Trakhtenbrot and Schützenberger have inspired a wealth of extensions as well as further research, and also led to recent practical applications, e.g. in verification of finite-state programs, in digital image compression and in speech-to-text processing (cf. Droste and Gastin [6, page 69]). In 2005, Droste and Gastin [5], [6], [10] established a quantitative extension of the Büchi–Elgot–Trakhtenbrot Theorem by introducing a weighted version of monadic second-order logic. More precisely, they could prove that, under certain assumptions, particular restrictions of their weighted logic are expressively equivalent to weighted automata. From a theoretical point of view, the logical description by means of classical as well as weighted monadic second-order formulas is particularly interesting. The formalism of monadic second-order logic can be applied for the abstract investigation of languages and formal power series, respectively, in order to derive theoretical results. On the other hand, formulas can be used to formalize qualitative as well as quantitative requirements or properties of systems.

As the central concepts under consideration are weighted automata, formal power series and weighted monadic second-order logic, the topic of this work is situated at the interface between Theoretical Computer Science, Algebra and Logic or, more generally, Model Theory. The main objective of this work is to "build bridges", i.e. to establish connections between the different notions and formalisms. On the one hand, we draw connections between the classical results from Language Theory and

---

[3] A self-contained proof of this generalized version of the Büchi–Elgot–Trakhtenbrot Theorem has not been available in the literature. The proof presented in this work is based on a suggestion of Manfred Droste, for which we are very grateful.

the ones concerning formal power series. On the other hand, we study the expressive power of the different approaches for the specification of languages and formal power series, respectively, and relate them to each other. More precisely, we derive that all approaches are expressively equivalent. Graphically, the main results in this work can be summarized in the following diagram:

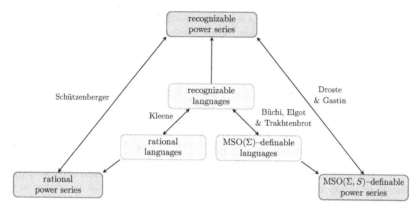

Next, we give a short overview by outlining the contents of this work. More detailed introductions are given at the beginning of the respective chapters and sections.

In Chapter 2, we start by gathering some general preliminaries on words and languages, which are the underlying structures for the entirety of this work. Furthermore, we consider the three above-mentioned approaches for the specification of languages: finite automata, rational operations and monadic second-order logic. We assume the reader to be familiar with the basics of Mathematical Logic, and particularly of Model Theory. However, Appendix A.1 provides a detailed introduction to the fundamental model-theoretic notions. Our treatment of monadic second-order logic in the context of words and languages is complemented by Appendix A.2. We refer to the corresponding section of Appendix A whenever it is convenient. In Theorem 2.3.6 we present the classical result of Kleene [18] showing that the sets of recognizable and rational languages are identical. Theorem 2.5.1 establishes the expressive equivalence of automata and monadic second-order logic, which is due to Büchi [4], Elgot [15] and Trakhtenbrot [43]. Moreover, we derive an extension of the Büchi–Elgot–Trakhtenbrot Theorem to formulas in Theorem 2.5.19. The corresponding proof presented at the end of Chapter 2 is a new contribution of this work[4].

The language-theoretic notions, methods and results are then used as starting point for our investigations in the subsequent chapters. At each point, we attempt to relate the respective extensions to their language-theoretic counterpart.

---

[4]A self-contained proof of this generalized version of the Büchi–Elgot–Trakhtenbrot Theorem has not been available in the literature. The proof presented in this work is based on a suggestion of Manfred Droste, for which we are very grateful.

Our study of weighted automata starts in Chapter 3. More precisely, we first introduce the fundamental theory of semirings and formal power series. We present several particular semirings that will occur throughout the whole work and serve as examples of possible weight structures. We then formally introduce the notion of a weighted automaton and its behavior. As we will demonstrate, these quantitative notions can be regarded as natural generalizations of finite automata and recognizable languages, respectively. However, we also shed light on some differences. Finally, we represent weighted automata in terms of matrices. These linear representations provide a further, more algebraic approach for the specification of formal power series. We prove that this approach even has the same expressive power as weighted automata.

Chapter 4 is devoted to the investigation of rational formal power series. To this end, we first define operations on formal power series having language-theoretic operations as natural counterparts. We further show that these operations yield a natural generalization of rational languages in the realm of formal power series. In particular, we derive an extension of Kleene's Theorem, which is due to Schützenberger [37] (see Theorem 4.3.1). More precisely, we first proceed by structural induction to show that every rational power series can be represented as the behavior of a weighted automaton. We then exploit interconnections between rational power series and linear systems of equations to prove, conversely, that the behavior of any weighted automaton is a rational power series.

The main subject of Chapter 5 is the logical examination of weighted automata and their behaviors. We introduce a weighted version of monadic second-order logic, which is due to Droste and Gastin [5]. By employing the formalism of weighted monadic second-order formulas, we obtain another method for the specification or representation of formal power series. However, we will demonstrate by the help of examples that the semantics of weighted formulas are in general more expressive than the behavior of weighted automata. Therefore, we restrict the weighted logic and introduce several fragments of it. By outlining the main steps of Droste and Gastin [6], we then derive that for commutative semirings the behaviors of weighted automata are precisely the power series definable by restricted weighted sentences (see Theorem 5.2.1). Ultimately, we introduce the notion of locally finite semirings, and we show that for this large class of semirings also unrestricted sentences define recognizable power series. Hence, locally finite commutative semirings admit an "unrestricted" generalization of the Büchi–Elgot–Trakhtenbrot Theorem (see Theorem 5.2.20).

Finally, in Chapter 6 we collect the main results and list the correspondences we establish throughout this work. Moreover, we indicate several research directions that tie into the theory presented in this work. These guide towards further work.

This work is intended to composed in a largely self-contained manner. Thus, we try to give a complete treatment of the included concepts as far as they are used in this work. Apart from the basics of Model Theory, which are presented in Appendix A, no particular previous knowledge from graduate level Mathematics and Computer Science is assumed. Whenever further notions or results are needed, they will be fully described or at least referenced. Furthermore, in order to illustrate the expressive power and diversity of languages, formal power series, and their respective representation tools, for each topic we present various examples. We hope that these fulfill the purpose of pointing out possible application scenarios of the theory.

For the rest of this work, we let $\mathbb{N} = \{0, 1, 2, 3, \dots\}$ **be the set of natural numbers and we denote by** $\mathbb{N}^+ := \mathbb{N} \setminus \{0\}$ **the set of positive natural numbers.**

2. Languages, Automata and Monadic Second-Order Logic

Although the main subject of this work is to examine formal power series, weighted automata and weighted monadic second-order logic, we start by considering the classical notions and results in the context of languages, finite automata and monadic second-order logic. These classical formalisms are the starting point of those in the weighted setting that will be considered in the subsequent chapters. Throughout this chapter, we further establish a clear overview over the classical results from Theoretical Computer Science, whose extensions and generalizations we will derive in the subsequent chapters. Thus, this chapter forms the basis of the whole work, as it provides all necessary foundations and preliminaries. More detailed introductions are given at the beginning of each section.

In Section 2.1, we introduce the most fundamental notions from Formal Language Theory. Section 2.2 is devoted to the study of finite automata and recognizable languages. In particular, we present a list of several decidability properties of finite automata and recognizable languages, respectively. In Section 2.3, we treat rational languages, prove several closure properties of recognizable languages and state Kleene's Theorem. In order to state and prove the Büchi–Elgot–Trakhtenbrot Theorem in Section 2.5, we study monadic second-order logic and its connection to words and languages in Section 2.4. We conclude this chapter with a Büchi–Elgot–Trakhtenbrot type result for formulas and not just for sentences. As a complete proof for this result has not been available in the literature, the proof presented at the end of Section 2.5 is a new contribution of this work.

2.1. Words and Formal Languages

> "Formal language theory is—together with automata theory (which is really inseparable from language theory)—the oldest branch of theoretical computer science. In some sense, the role of language and automata theory in computer science is analogous to that of philosophy in general science: it constitutes the stem from which the individual branches of knowledge emerge."
>
> Salomaa [36, page 105]

This section gathers the fundamental structures and preliminary notions in the context of Formal Language Theory that will be used throughout this work. One could say that words and languages are the playthings in this chapter, as the subsequent sections are devoted to the introduction of modeling approaches to represent them. However, the classical notions and their properties, which are introduced in this section, also play a central role in the weighted setting, with which the subsequent chapters are concerned.

We set up notation and terminology mainly following Sakarovitch [33, Chapter 0] and Droste [11, § 1].

2.1.1 Definition. An **alphabet** is a non-empty finite set Σ. The elements of an alphabet are called **letters** or **symbols**.

2.1.2 Example. The Latin alphabet $\Sigma = \{a, A, b, B, c, C, \ldots, z, Z\}$ is probably the first alphabet one thinks of. However, elements of an alphabet may also be

- numbers, e.g. $\Sigma = \{0, 1\}$,

- words, e.g. $\Sigma = \{\texttt{bye}, \texttt{tschuess}, \texttt{ciao}, \texttt{farvel}\}$,

- or various symbols, e.g. $\Sigma = \{A, B, 1, 2, 3, 4, +, \times, \odot\}$.

In the upcoming examples, we usually work with (subsets of) the alphabet

$$\Sigma = \{a, b, c, d\}.$$

2.1.3 Definition. Let Σ be an alphabet.

a) A (finite) **word** over Σ is a finite sequence of the form

$$w = a_1 \ldots a_n$$

with $n \in \mathbb{N}$ and $a_1, \ldots, a_n \in \Sigma$. The word obtained for $n = 0$ is referred to as the **empty word** and is denoted by ε. Words are also called **strings**.

b) Let $w = a_1 \ldots a_n$ be a word over Σ. Then n is called the **length** of w and is denoted by $|w|$. In particular, we have $|\varepsilon| = 0$.

c) Given two words $w = a_1 \ldots a_n$ and $v = b_1 \ldots b_m$ over Σ, their **concatenation** is defined to be the word

$$w \cdot v := a_1 \ldots a_n b_1 \ldots b_m.$$

Thus, we obtain the concatenation of w and v by writing the word v immediately after the word w. Instead of $w \cdot v$ we usually write wv.

d) Given a word w over Σ, we define **powers** of w inductively by

$$w^0 := \varepsilon,$$
$$w^{n+1} := w^n \cdot w \quad (\text{for } n \in \mathbb{N}).$$

2.1.4 Remark.

a) Formally, a word over Σ is a tuple $w = (a_1, \ldots, a_n)$ with $n \in \mathbb{N}$ and $a_1, \ldots, a_n \in \Sigma$. In particular, two words $(a_1, \ldots, a_n)$ and $(b_1, \ldots, b_m)$ over Σ are equal if they have the same length $n = m$ and fulfill $a_i = b_i$ for any $i \in \{1, \ldots, n\}$. Thus, the equality of words depends on the underlying alphabet.

b) For convenience, we simply denote the tuple $(a_1, \dots, a_n)$ by the sequence $a_1 \dots a_n$. However, this notation does not work in general. For instance, if we consider the alphabet $\Sigma = \{a, aa\}$, then it is not clear whether the sequence $w = aaa$ stands for the word (aa, a), (a, aa), or (a, a, a). Therefore, we assume that every alphabet Σ in this work allows an unambiguous assignment of sequences to the letters in Σ they consist of. This assumption guarantees that every word in Σ^* has a unique written form as concatenation of letters in Σ.

2.1.5 Example. Consider the alphabet $\Sigma = \{\texttt{hello}, \texttt{my}, \texttt{friend}, \texttt{␣}, \texttt{!}, \texttt{?}\}$. Then

$$u = \texttt{hello?hello?}$$
$$v = \texttt{?!my␣}$$
$$w = \texttt{hello␣my␣friend!}$$

are words over Σ. This example shows that, over an appropriate alphabet, also sentences can be interpreted as words. We note further that the word u has length 4 with respect to the alphabet Σ, whereas it has length 12 with respect to the alphabet $\Sigma' = \{\texttt{h}, \texttt{e}, \texttt{l}, \texttt{o}, \texttt{?}\}$.

2.1.6 Definition. Let Σ be an alphabet. We set

$$\Sigma^* := \{a_1 \dots a_n \mid n \in \mathbb{N}, a_1, \dots, a_n \in \Sigma\},$$

i.e. Σ^* is the set containing all (finite) words over Σ. Moreover, we let

$$\Sigma^+ := \Sigma^* \setminus \{\varepsilon\}$$

be the set of all **non-empty** words over Σ. Given $n \in \mathbb{N}$, we denote by Σ^n the set of all words of length n over Σ.

We treat Σ as a subset of Σ^*, i.e. we do not distinguish between letters of Σ and words of length 1 over Σ. In particular, we have

$$\Sigma^* = \bigcup_{n \in \mathbb{N}} \Sigma^n \text{ and } \Sigma^+ = \bigcup_{n \in \mathbb{N}^+} \Sigma^n,$$

with

$$\Sigma^0 = \{\varepsilon\}, \ \Sigma^1 = \Sigma, \ \Sigma^2 = \{ab \mid a, b \in \Sigma\}, \text{ etc.}$$

We note further that the concatenation of words provides a binary operation $\cdot$ on Σ^*.

2.1.7 Example.

a) For the singleton alphabet $\Sigma = \{a\}$ we obtain

$$\Sigma^* = \{\varepsilon, a, aa, aaa, \dots\} = \{a^n \mid n \in \mathbb{N}\},$$
$$\Sigma^+ = \{a^n \mid n \in \mathbb{N}^+\},$$
$$\Sigma^n = \{a^n\} \quad (\text{for } n \in \mathbb{N}).$$

b) For the alphabet $\Sigma = \{a, b\}$ we obtain

$$\Sigma^* = \{\varepsilon, a, b, aa, ab, ba, bb, aaa, aab, aba, abb, baa, bab, bba, bbb, \dots\}.$$

To further study Σ^* and its properties, we recall some algebraic notions in the following.

2.1.8 Definition.

a) A **monoid** is a triple $(M, \cdot, 1)$ consisting of a non-empty set M, an associative binary operation $\cdot \colon M \times M \to M$, $(m_1, m_2) \mapsto m_1 \cdot m_2 := \cdot(m_1, m_2)$ and a neutral element $1 \in M$ with $m \cdot 1 = 1 \cdot m = m$ for any $m \in M$. For convenience, the tuple $(M, \cdot, 1)$ is simply denoted by M if there is no confusion likely to arise. Further, we often omit the symbol $\cdot$ of the binary operation and write $m_1 m_2$ instead of $m_1 \cdot m_2$ for elements $m_1, m_2 \in M$.

b) A monoid M is called **commutative** if $m_1 \cdot m_2 = m_2 \cdot m_1$ holds for any $m_1, m_2 \in M$.

It is well-known that the neutral element of a monoid M is unique. Indeed, if we are given neutral elements 1 and $1'$ of M, then we obtain $1 = 1 \cdot 1' = 1'$.

2.1.9 Example. A typical example of a monoid are the natural numbers $(\mathbb{N}, +, 0)$ where $+$ denotes the usual addition on $\mathbb{N}$, which is even a commutative operation.

2.1.10 Definition. Let $(M, \cdot, 1)$ and $(M', \circ, 1')$ be monoids.

a) A **monoid homomorphism** from M into M' is a map $h \colon M \to M'$ fulfilling

- $h(1) = 1'$ and
- $h(m_1 \cdot m_2) = h(m_1) \circ h(m_2)$ for any $m_1, m_2 \in M$.

b) A bijective monoid homomorphism is called a **monoid isomorphism**.

c) The monoids M and M' are called **isomorphic** if there exists a monoid isomorphism from M into M'.

2.1.11 Example. Let Σ be an alphabet.

a) The most important type of a monoid in this work is $(\Sigma^*, \cdot, \varepsilon)$, called the **free monoid generated by the alphabet** Σ. Indeed, the concatenation of words over Σ is an associative operation whose neutral element is given by the empty word. One can even show that $(\Sigma^*, \cdot, \varepsilon)$ is the smallest monoid that contains Σ and is closed under concatenation (cf. Eilenberg [14, page 5]).

b) The monoid Σ^* is commutative if and only if the underlying alphabet Σ contains just a single letter. In fact, if Σ contains two distinct letters $a \neq b$, then the inequality $a \cdot b \neq b \cdot a$ implies that Σ^* is not commutative. For the converse, we assume that $\Sigma = \{a\}$ is a singleton. The free monoid Σ^* is then isomorphic to the commutative monoid $(\mathbb{N}, +, 0)$ via the monoid isomorphism $\mathbb{N} \to \Sigma^*$, $n \mapsto a^n$, whose inverse is given by the length function $\Sigma^* \to \mathbb{N}, a^n \mapsto |a^n| = n$. Therefore, $(\mathbb{N}, +, 0)$ is often said to be a *free* monoid as well.

The reason for calling monoids generated by some alphabet *free* monoids is the following *universal property* (cf. Sakarovitch [33, page 24]), which we will apply various times throughout this work.

2.1.12 Lemma. *Let Σ be an alphabet and $(M, \circ, 1)$ a monoid. Then every map $h \colon \Sigma \to M$ can be uniquely extended to a monoid homomorphism $\hat{h}$ from the free monoid $(\Sigma^*, \cdot, \varepsilon)$ into the monoid $(M, \circ, 1)$. In particular, every monoid homomorphism h from the free monoid Σ^* into a monoid M is uniquely determined by its restriction $h\big|_\Sigma$ to the underlying alphabet Σ.*

Proof. We put $\hat{h}(\varepsilon) = 1$ and $\hat{h}(a_1 \dots a_n) = h(a_1) \circ \dots \circ h(a_n)$ for any $n \in \mathbb{N}^+$ and any $a_1, \dots, a_n \in \Sigma$. Clearly, the map $\hat{h}$ extends h and is a monoid homomorphism by definition. To prove the uniqueness of $\hat{h}$, let $\tilde{h}$ be another monoid homomorphism from the free monoid $(\Sigma^*, \cdot, \varepsilon)$ into the monoid $(M, \circ, 1)$ extending h. Then for each word $w = a_1 \dots a_n \in \Sigma^*$ we obtain

$$\tilde{h}(w) = \tilde{h}(a_1 \dots a_n)$$
$$= \tilde{h}(a_1) \circ \dots \circ \tilde{h}(a_n)$$
$$= h(a_1) \circ \dots \circ h(a_n)$$
$$= \hat{h}(w). \qquad \square$$

We denote the above described (unique) extension $\hat{h}$ of a map h again by h if no confusion is likely to arise.

2.1.13 Example. Consider the constant map $h \colon \Sigma \to \mathbb{N}$, $a \mapsto 1$. It is easy to see that its extension $\hat{h}$ is precisely the length function $|\cdot| \colon \Sigma^* \to \mathbb{N}$, $w \mapsto |w|$. In other words, the length function is the unique monoid homomorphism from $(\Sigma^*, \cdot, \varepsilon)$ into $(\mathbb{N}, +, 0)$ that maps each letter of Σ to 1. In particular, we have

$$|w \cdot v| = |w| + |v| \text{ and } |w^n| = n \cdot |w|$$

for each $w, v \in \Sigma^*$ and $n \in \mathbb{N}$.

Usually, the monoid homomorphisms we deal with are of the form $h \colon \Sigma^* \to \Gamma^*$, where both Σ and Γ are alphabets. Such maps allow us to change the underlying alphabet, as they assign words over Γ to words over Σ. Based on Droste and Kuich [9, page 16 f.], we now introduce some properties of such monoid homomorphisms.

2.1.14 Definition. Let Σ, Γ be alphabets and $h \colon \Sigma^* \to \Gamma^*$ a monoid homomorphism. We call h

- **non-extending** if $|h(w)| \leq |w|$ for any $w \in \Sigma^*$.

- **non-deleting** if $|h(w)| \geq |w|$ for any $w \in \Sigma^*$.

- **length-preserving** if $|h(w)| = |w|$ for any $w \in \Sigma^*$.

2.1.15 Remark. Let Σ, Γ be alphabets and $h \colon \Sigma^* \to \Gamma^*$ a monoid homomorphism. Then the image $h(w)$ of a word w is uniquely determined by the images of the letters it consists of (see Lemma 2.1.12). Moreover, we know from Example 2.1.13 that the length function is additive. Therefore, we obtain the following equivalences for non-extending homomorphisms:

$$h \text{ is non-extending} \Leftrightarrow |h(w)| \leq |w| \text{ for any } w \in \Sigma^*$$
$$\Leftrightarrow |h(a)| \leq |a| = 1 \text{ for any } a \in \Sigma$$
$$\Leftrightarrow h(a) \in \{\varepsilon\} \cup \Gamma \text{ for any } a \in \Sigma$$
$$\Leftrightarrow h(\Sigma) \subseteq \{\varepsilon\} \cup \Gamma$$
$$\Leftrightarrow h \text{ is the extension of a map } \Sigma \to \{\varepsilon\} \cup \Gamma \text{ (see Lemma 2.1.12)}$$

For non-deleting homomorphisms we obtain the equivalences:

$$h \text{ is non-deleting} \Leftrightarrow |h(w)| \geq |w| \text{ for any } w \in \Sigma^*$$
$$\Leftrightarrow |h(a)| \geq |a| = 1 \text{ for any } a \in \Sigma$$
$$\Leftrightarrow h(a) \neq \varepsilon \text{ for any } a \in \Sigma$$
$$\Leftrightarrow h^{-1}(\varepsilon) = \{\varepsilon\}$$
$$\Leftrightarrow h \text{ is the extension of a map } \Sigma \to \Gamma^+ \text{ (see Lemma 2.1.12)}$$

Combining the two observations above, we further achieve:

$$h \text{ is length-preserving} \Leftrightarrow |h(w)| = |w| \text{ for any } w \in \Sigma^*$$
$$\Leftrightarrow h \text{ is non-extending and non-deleting}$$
$$\Leftrightarrow h(\Sigma) \subseteq \Gamma, \text{i.e. } h \text{ maps letters to letters}$$
$$\Leftrightarrow h \text{ is the extension of a map } \Sigma \to \Gamma \text{ (see Lemma 2.1.12)}$$

Length-preserving homomorphisms are also referred to as strictly alphabetic homomorphisms or codings (cf. Sakarovitch [33, page 25] and Salomaa and Soittola [35, page 4 f.])

To conclude this section, we now introduce the notion of a (formal) language and define several language-theoretic operations.

2.1.16 Definition. Let Σ be an alphabet. We call each subset L of Σ^* a **(formal) language** over Σ.

2.1.17 Example. Consider the alphabet $\Sigma = \{a, b\}$.

a) The language

$$L_{ab} = \{w \in \Sigma^* \mid w \text{ ends with } ab\} = \{vab \mid v \in \Sigma^*\}$$

will occur various times throughout this chapter.

b) Let $|w|_a$ denote the number of occurrences of the letter $a \in \Sigma$ in a word w over Σ. Then
$$L_{\mathrm{mod}} = \{w \in \Sigma^* \mid |w|_a \equiv 1 \,(\mathrm{mod}\ 3)\}$$
is a language over Σ.

c) Another language over Σ is given by
$$L_{\mathrm{pump}} = \{a^n b^n \mid n \in \mathbb{N}\}.$$

2.1.18 Definition. Let L_1, L_2 and L be languages over an alphabet Σ. We define

- the **concatenation**, or the **product**, of L_1 and L_2 by
$$L_1 \cdot L_2 := L_1 L_2 := \{wv \mid w \in L_1, v \in L_2\},$$

- **powers** of L by
$$L^0 := \{\varepsilon\},$$
$$L^{n+1} := L^n \cdot L \quad (\text{for } n \in \mathbb{N}),$$

- the **(Kleene) star** of L by
$$L^* := \bigcup_{n \in \mathbb{N}} L^n,$$

- the **(Kleene) plus** of L by
$$L^+ := \bigcup_{n \in \mathbb{N}^+} L^n.$$

2.1.19 Remark. Let L be a language over some alphabet Σ.

a) The empty language annihilates every language, i.e. $L \cdot \emptyset = \emptyset \cdot L = \emptyset$.

b) The singleton language $\{\varepsilon\}$ acts as the neutral element of the language-theoretic concatenation. More precisely, the triple $(\mathcal{P}(\Sigma^*), \cdot, \{\varepsilon\})$, where we denote by $\mathcal{P}(A)$ the power set of a set A, constitutes a monoid that in general is not commutative. In particular, we have $\emptyset^0 = \{\varepsilon\}$ and thus $\emptyset^* = \{\varepsilon\}$.

c) One can easily verify that L^* is the smallest superset of L that contains the empty word and is closed under concatenation. Moreover, it requires just a simple computation to show that the Kleene star satisfies $(L^*)^* = L^*$, i.e. the Kleene star is an idempotent unary operator. The Kleene plus and the Kleene star of L coincide if and only if L contains the empty word, i.e.
$$L^* = L^+ \ \Leftrightarrow \ \varepsilon \in L.$$

d) The language-theoretic operations which we just introduced are consistent with our notations for alphabets from Definition 2.1.6. More precisely, any alphabet Σ can be regarded as a language over itself, since we treat Σ as subset of Σ^*.

2.1.20 Example. Using the language-theoretic operations from Definition 2.1.18, the language from Example 2.1.17a) can be rewritten as

$$L_{ab} = \{w \in \Sigma^* \mid w \text{ ends with } ab\} = \{vab \mid v \in \Sigma^*\}$$
$$= \Sigma^*\{ab\} = \Sigma^*\{a\}\{b\}.$$

In the subsequent sections, we introduce modeling approaches each of which captures an important class of languages. Moreover, we show that these approaches are all equivalent in the sense that they all capture the same class of languages.

2.2. Finite Automata

The finite automata we consider in this work began to be studied in the 1950s, motivated in part by a practical interest – the design of sequential logical circuits – and a more speculative one – the modeling of human neural activity (cf. Straubing [41, page 7]). Since the origins, the theory of finite automata has developed, stimulated both by its possible applications and by its inner mathematical orientations (cf. Perrin [27, page 3]). From the mathematical point of view, one important application of finite automata is concerned with decidability problems. In this work, automata are of particular interest, as they provide a finite tool to model languages that may themselves be infinite. More precisely, they provide testing procedures or mechanisms for words, respectively. Languages that can be modeled by a finite automaton are called recognizable. In this section, we develop the fundamental theory of finite automata and recognizable languages. In particular, in Theorem 2.2.14 we present several decidability properties of finite automata and recognizable languages, respectively.

The primary source for this section is Perrin [27, § 2]. However, we also took inspiration from Khoussainov and Nerode [17, § 2.2] as well as Droste [11, § 1].

Before we give all the formal definitions, we first want to informally describe the concept of a finite automaton and its connection to words. A finite automaton can be viewed as a "black box" that either accepts or rejects a given word w. Pictorially, we have:

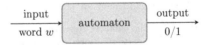

The output 1 means that the automaton accepts the word w and the output 0 represents the rejection of the word w by the automaton. More precisely, a finite automaton consists of finitely many states. The arrival of a letter may lead to a change of the state. Thus, given a word w, we can "run" through a state sequence in the automaton by reading the letters in w and changing the states appropriately.

Throughout the rest of this work, let Σ be an alphabet.

2.2.1 Definition. A **finite automaton** over Σ is a quadruple $\mathcal{A} = (Q, I, T, F)$ consisting of

- a non-empty finite set Q of **states**,

- subsets $I, F \subseteq Q$ whose elements are called **initial states** and **final states**, respectively, and

- a subset $T \subseteq Q \times \Sigma \times Q$ whose elements are called **transitions**.

The condition $(p, a, q) \in T$ means: If the automaton $\mathcal{A}$ is in state p, then it can change into state q by reading the letter a.

Later, we will call such automata *classical* to stress the difference between these and the *weighted* automata which we will introduce in Chapter 3. Moreover, we often omit the adjective *finite* when we refer to finite automata.

2.2.2 Definition. Let $\mathcal{A} = (Q, I, T, F)$ be a finite automaton over Σ. We call $\mathcal{A}$

- **deterministic** if $|I| = 1$, i.e. there exists a unique initial state, and the transition set fulfills the following condition

$$\forall (p, a, q), (p, a, q') \in T : q = q'$$

  i.e. successor states are unique.

- **complete** if it satisfies the condition

$$\forall p \in Q \; \forall a \in \Sigma \; \exists q \in Q : (p, a, q) \in T$$

  which ensures the existence of successor states.

We refer to an automaton that is deterministic and complete as complete deterministic automaton. Many other authors, e.g. Khoussainov and Nerode [17, § 2.2] and Straubing [41, § I.2], refer to such automata simply as deterministic automata.

2.2.3 Remark. Given a finite automaton $\mathcal{A} = (Q, I, T, F)$ over Σ, we obtain an equivalent description of the automaton if we replace the transition set T by the **transition function**, which is given by

$$\delta \colon Q \times \Sigma \to \mathcal{P}(Q),$$
$$(p, a) \mapsto \{q \in Q \mid (p, a, q) \in T\}.$$

Since we do not require $\mathcal{A}$ to be complete, the set of successor states $\delta(p, a)$ of a given pair $(p, a) \in Q \times \Sigma$ is possibly empty. However, if $\mathcal{A}$ is a complete and deterministic, then every pair $(p, a) \in Q \times \Sigma$ has a unique successor state $q \in Q$ such that $(p, a, q) \in T$, i.e. the transition function can be identified with $\delta \colon Q \times \Sigma \to Q$ defined by

$$\delta(p, a) = q \; :\Leftrightarrow \; (p, a, q) \in T$$

for $p, q \in Q$ and $a \in \Sigma$.

Usually, we illustrate a finite automaton $\mathcal{A} = (Q, I, T, F)$ by its **state diagram**, which is a directed graph with labeled edges. More precisely, we have the following:

- The vertices of the state diagram are given by the states of the automaton. Pictorially, each state $q \in Q$ is represented by a circle with label q:

- Each vertex that corresponds to an initial state is equipped with an incoming arrow, i.e. each $q \in I$ is depicted as:

- Likewise, each vertex corresponding to a final state is equipped with an outgoing arrow. i.e. each $q \in F$ is depicted as:

- The edges of the state diagram are determined by the transitions of the automaton. More precisely, each transition $(p, a, q) \in T$ with $p \neq q$ is represented by an edge directed from vertex p to vertex q that is labeled by the letter a. Pictorially, this means:

If the states p and q of the transition coincide, then we equip the vertex p with a loop that is labeled by the letter a. Pictorially, this means:

If there are several transitions from state p to state q, then the several edges may be replaced by a single edge carrying several labels.

Given such a state diagram, we can conversely specify the formal definition of the corresponding automaton.

2.2.4 Example. The state diagram

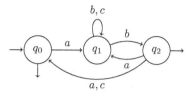

corresponds to the automaton $\mathcal{A} = (Q, I, T, F)$ over $\Sigma = \{a, b, c\}$ with

$Q = \{q_0, q_1, q_2\}$,

$I = \{q_0\}$,

$T = \{(q_0, a, q_1), (q_1, b, q_1), (q_1, b, q_2), (q_1, c, q_1), (q_2, a, q_0), (q_2, a, q_1), (q_2, c, q_0)\}$,

$F = \{q_0, q_2\}$.

This automaton is not deterministic, since e.g. the successor state of the pair (q_1, b) is not unique. More precisely, we have $\delta(q_1, b) = \{q_1, q_2\}$ using the transition function from Remark 2.2.3. Moreover, the automaton is not complete, since e.g. there is no successor state for the pair (q_0, b).

2.2.5 Example. Finite automata can model simple machines. If we consider the alphabet $\Sigma = \{\texttt{pay}, \texttt{cancel}, \texttt{choose}, \texttt{remove}\}$, then the (deterministic) automaton given by the following state diagram represents the (simplified) working method of a snack machine:

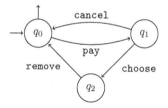

The elements of Σ represent the possible actions of the customer, and thus any word over Σ can be understood as a sequence of actions. The states of the automaton model the following three states of the snack machine: "waiting for a customer" (q_0), "waiting for the customer's choice" (q_1) and "giving out the respective snack" (q_2). The transition $(q_1, \texttt{cancel}, q_0)$, for instance, corresponds to the customer's cancellation of his or her purchase (after the payment). Intuitively, each valid purchase has to start with the payment followed by the choice of the snack, and is completed by the removal of the respective snack. If the costumer has paid but not yet chosen a snack, then he or she can also cancel the purchase. The only final state, which is at the same time the only initial state, is reached when the purchase is either completed or canceled.

As we have already mentioned at the beginning of this section, finite automata provide a tool to model languages. We now formally define this connection between finite automata and languages. In particular, we introduce the testing mechanism for words provided by a finite automaton.

2.2.6 Definition. Let $\mathcal{A} = (Q, I, T, F)$ be a finite automaton over Σ.

a) A **path**, or a **run**, in $\mathcal{A}$ is an alternating sequence $P = q_0 a_1 q_1 \ldots a_n q_n \in Q(\Sigma Q)^*$ with $n \in \mathbb{N}$ such that $t_i = (q_{i-1}, a_i, q_i) \in T$ for $i = 1, \ldots, n$, i.e. P can be regarded as a sequence of transitions. We represent the path P pictorially by

$$q_0 \xrightarrow{a_1} q_1 \xrightarrow{a_2} q_2 \to \cdots \to q_{n-1} \xrightarrow{a_n} q_n.$$

The word $a_1 \ldots a_n \in \Sigma^*$ is called **label** of the run P. We call the run P **successful** if $q_0 \in I$ and $q_n \in F$.

b) We say that $\mathcal{A}$ **accepts** or **recognizes** a word $w \in \Sigma^*$ if there exists a succesful path in $\mathcal{A}$ with label w. The **language recognized by the automaton** $\mathcal{A}$ is defined by

$$L(\mathcal{A}) := \{w \in \Sigma^* \mid w \text{ is recognized by } \mathcal{A}\}.$$

Hence, $L(\mathcal{A})$ contains precisely the labels of successful paths in $\mathcal{A}$.

The *testing procedure* of a finite automaton for a given word w can be described as follows: First, the automaton arbitrarily chooses an initial state where it starts. Then it successively consumes the letters of w from front to back. For each letter, it arbitrarily chooses one of the applicable transitions and by this changes to a new state. If a final state is reached after processing all letters of the word w, then it is accepted by the automaton. If there is no such run that consumes the word w and leads to a final state, then the word is rejected. Thus, a finite automaton, in general, causes a non-deterministic procedure. Correspondingly, finite automata are often referred to as non-deterministic finite automata[1] (cf. Khoussainov and Nerode [17, § 2.2]). On the other hand, for a deterministic finite automaton the above described procedure is deterministic in the following sense: First, there is only one initial state where the automaton can start. Then for each letter there is at most one applicable transition. Hence, there is at most one successful run for a given word.

2.2.7 Definition. We call a language L over Σ **recognizable** if there is a finite automaton $\mathcal{A}$ such that $L = L(\mathcal{A})$.

2.2.8 Remark. We have seen, e.g. in Example 2.2.4, that not every finite automaton is deterministic and complete. However, complete deterministic automata have the same expressive power as the general ones, i.e. a language is recognizable if and only if it is recognized by some complete deterministic automaton.

---

[1] Adapting this terminology, on the one hand there are non-deterministic automata that are not deterministic, but on the other hand there are also non-deterministic automata that are deterministic (see Definition 2.2.2). Thus, the adjective *non-deterministic* may be misleading, and for this reason we omit it and just speak of finite automata.

More precisely, using the so-called *power set construction* (cf. Droste [11, Theorem 1.5]), each automaton $\mathcal{A}$ can be effectively transformed into a complete deterministic automaton $\mathcal{A}'$ such that $L(\mathcal{A}) = L(\mathcal{A}')$.

2.2.9 Example. Consider the alphabet $\Sigma = \{a, b\}$.

a) The automaton $\mathcal{A}$ represented by the state diagram

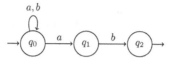

recognizes the language $L_{ab} = \{w \in \Sigma^* \mid w$ ends with $ab\}$. Indeed, every successful path in $\mathcal{A}$ has to start in the initial state q_0 and has to end in the final state q_2. To reach state q_2 starting from state q_0, the path needs to lead through state q_1. Therefore, the label of any successful path ends with ab, i.e. we have $L(\mathcal{A}) \subseteq L_{ab}$. Viceversa, any word $w \in L_{ab}$ can be written as $w = vab$ for some $v \in \Sigma^*$. The word v can then be realized using the loop in state q_0. After that, the automaton first changes to state q_1 and then to state q_2 by successively reading the letters a and b. This procedure represents a successful run with label w. Hence, we obtain the inclusion $L_{ab} \subseteq L(\mathcal{A})$. Furthermore, the automaton $\mathcal{A}$ is not deterministic and not complete, since we have e.g. $\delta(q_0, a) = \{q_0, q_1\}$ and $\delta(q_1, a) = \emptyset$ for the transition function from Remark 2.2.3. However, the state diagram

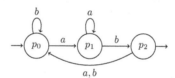

represents a complete deterministic automaton $\mathcal{A}'$ such that $L(\mathcal{A}') = L(\mathcal{A}) = L_{ab}$. One obtains the automaton $\mathcal{A}'$ by applying the power set construction (see Remark 2.2.8) to the automaton $\mathcal{A}$ (and by omitting states that cannot be reached from the initial state).

b) The complete deterministic automaton given by the state diagram

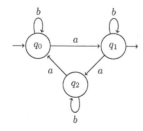

recognizes the language $L_{\mathrm{mod}} = \{w \in \Sigma^* \mid |w|_a \equiv 1 \,(\mathrm{mod}\,3)\}$. Intuitively, we have a correspondence between the state q_i and the language $\{w \in \Sigma^* \mid |w|_a \equiv i \,(\mathrm{mod}\,3)\}$, for each $i \in \{0, 1, 2\}$. More precisely, given a word $w \in \Sigma^*$ with $|w|_a \equiv i \,(\mathrm{mod}\,3)$, the unique path with label w starting in the initial state q_0 ends in state q_i.

In particular, the languages L_{ab} and L_{mod} are recognizable.

In the following, we present several properties of recognizable languages and finite automata, respectively. The subsequent propositions show that monoid homomomorphisms of the form $h\colon \Sigma^* \to \Gamma^*$, where Σ and Γ are alphabets, are compatible with the concept of recognizability. More precisely, recognizability is preserved both by taking preimages and images of languages under h. Proofs for these results can be found in Hopcroft, Motwani and Ullman [16, § 4.2.3 and § 4.2.4] and in Eilenberg [14, page 19].

2.2.10 Proposition. *Let Σ, Γ be alphabets and $h\colon \Sigma^* \to \Gamma^*$ a monoid homomorphism. If L is a recognizable language over Γ, then its preimage $h^{-1}(L)$ is a recognizable language over Σ.*

2.2.11 Proposition. *Let Σ, Γ be alphabets and $h\colon \Sigma^* \to \Gamma^*$ a non-deleting monoid homomorphism. If L is a recognizable language over Σ, then its image $h(L)$ is a recognizable language over Γ.*

The next result, commonly referred to as *Pumping Lemma*, describes an essential property of recognizable languages. It was first proved by Rabin and Scott [31, Lemma 8] in 1959.[2] Their motivation was to establish the decidability of the *Infinity Problem* (see Theorem 2.2.14). However, the Pumping Lemma also provides a useful tool to disprove the recognizability of a specific language in question. We use a slightly different wording of the original result, which is proved in Hopcroft, Motwani and Ullman [16, § 4.1.1].

2.2.12 Proposition (Pumping Lemma). *Let L be a recognizable language over Σ. Then there exists a natural number $n \in \mathbb{N}^+$ such that each word $w \in L$ with length $|w| \geq n$ can be written as $w = xyz$ with $x, y, z \in \Sigma^*$ satisfying the following conditions:*

(i) $|y| \geq 1$,

(ii) $|xy| \leq n$,

(iii) $xy^pz \in L$ for any $p \in \mathbb{N}$.

2.2.13 Example. We claim that the language $L_{\mathrm{pump}} = \{a^n b^n \mid n \in \mathbb{N}\}$ is not recognizable. To prove this, we apply Proposition 2.2.12; more precisely we apply its contrapositive. Thus, let $n \in \mathbb{N}$ be any natural number and consider the word $w = a^n b^n \in L_{\mathrm{pump}}$ with length $|w| = 2n \geq n$. If we write $w = xyz$ with $x, y, z \in \Sigma^*$ such that conditions (i) and (ii) are fulfilled, then it must be the case that $xy = a^m$ for $m = |xy| \leq n$. In particular, we obtain $y = a^k$ for $k = |y| \geq 1$ and thus e.g. the word $xy^0z = a^{n-k}b^n$ is not contained in L_{pump}. By this, condition (iii) is violated, and hence L_{pump} cannot be recognizable.

---

[2]In [31], Rabin and Scott provided the first systematic treatment of the theory of finite automata and they presented virtually everything known up to 1959 (cf. Eilenberg [14, page 74]).

Finally, before we turn to another formalism to describe languages, we want to mention the following decidability results concerning automata and recognizable languages, respectively. Detailed proofs of these can be found in Droste [11, Theorem 1.13], Hopcroft, Motwani and Ullman [16, § 4.3], Rabin and Scott [31, Corollary 7.1, Corollary 9.1 and Corollary 10.1].

2.2.14 Theorem. *Let A and A' be finite automata over Σ. Then the following problems are decidable*[3]*:*

a) *Given a word w over Σ, is w recognized by A?*
(Word Problem)

b) *Is the language recognized by A empty?*
(Emptiness Problem)

c) *Does A recognize all words over Σ?*
(Universality Problem)

d) *Does A recognize infinitely many words?*
(Infinity Problem)

e) *Is the language $L(A)$ contained in the language $L(A')$?*
(Inclusion Problem)

f) *Do A and A' recognize the same language?*
(Equivalence Problem)

In the subsequent sections, we introduce further formalisms that provide modeling approaches for languages. Moreover, we show that each of these is equivalent to the one of finite automata. Thus, we obtain multiple equivalent tools to specify languages.

2.3. Kleene's Theorem

This section is concerned with the study of rational languages and the examination of their relation to finite automata and recognizable languages, respectively. A language is called rational if it can be constructed from finitely many singleton languages by applying the language-theoretic operations union, concatenation and Kleene star. The notion of rational languages – commonly also referred to as regular languages – was first introduced in the seminal work of Kleene [18] in 1956. Kleene is even considered one of the founders of Theoretical Computer Science, especially of Formal Language and Automata Theory. Indeed, he was the first to formalize the early attempts of McCulloch and Pitts [24] to capture the notion *finite-state machines.* McCulloch's and Pitt's work was mainly motivated by the logical modeling and the behavior of (biological) neural networks and nervous systems, respectively. Further details on the

---

[3]A problem is **decidable** if it is solvable by an algorithm, i.e. there exists an effective procedure to decide the problem

historical developments and application domains of Automata Theory can be found in Perrin [27, page 3 f.]. The fundamental result of Theoretical Computer Science showing that the language-theoretic notions of recognizability and rationality are equivalent is also due to Kleene [18], and is therefore commonly referred to as Kleene's Theorem. This equivalence of the two modeling approaches is the central result of this section. In Chapter 4 we will extend Kleene's formalism of rational languages and prove a Kleene type result for formal power series. This result is due to Schützenberger [37], contains Kleene's Theorem as a special case, and is therefore commonly referred to as Kleene–Schützenberger Theorem.

In this section, we mainly follow Khoussainov and Nerode [17, § 2.3 and § 2.5], Sakarovitch [33, § I.2.2], and we also took inspiration from Perrin [27, § 3 and § 4].

2.3.1 Definition. We define the set of **rational languages** over Σ inductively, as follows:

(1) The empty set $\emptyset$ is a rational language.

(2) For each $a \in \Sigma$, the singleton language $\{a\}$ is a rational language over Σ.

(3) If L_1 and L_2 are rational languages over Σ, then so is their union $L_1 \cup L_2$.

(4) If L_1 and L_2 are rational languages over Σ, then so is their concatenation $L_1 L_2$.

(5) If L is a rational language over Σ, then so is its Kleene star L^*.

Union, concatenation and Kleene star are called **rational operations**.

Rational languages are often specified using so-called *rational expressions* (cf. [27, § 3]). Furthermore, we want to mention that our terminology is not the common one. The terms *rational* and *recognizable* originate from Eilenberg [14]. In the literature, rational or recognizable languages are often referred to as *regular* languages (cf. [17, § 2.5] and Straubing [41, § I.2]). However, according to Straubing [41, page 8] the term *regular* is a "bit unfortunate" and still almost used universally, while Eilenberg's terminology is "better motivated" but "never really caught on". The main reason for us to follow Eilenberg is that the adjective *rational* emphasizes the analogy between rational languages and the rational formal power series which we will introduce in Chapter 4.

2.3.2 Remark. It follows directly from Definition 2.3.1 that every finite language is rational. Indeed, the singleton language $\{\varepsilon\} = \emptyset^*$ is rational by steps (1) and (5) from Definition 2.3.1. Since we further have $\{a_1 \ldots a_n\} = \{a_1\} \ldots \{a_n\}$ for any $n \in \mathbb{N}^+$ and $a_1, \ldots, a_n \in \Sigma$, steps (2) and (4) imply that every singleton language over Σ is rational. Now every finite language is a finite union of singleton languages and thus rational by step (3).

Hence, the set of rational languages over Σ is the smallest subset of $\mathcal{P}(\Sigma^*)$ that contains all finite languages and is closed under union, concatenation and Kleene star.

2.3.3 Example. Consider the alphabet $\Sigma = \{a, b\}$.

a) The rearrangements

$$L_{ab} = \{w \in \Sigma^* \mid w \text{ ends with } ab\} = \{vab \mid v \in \Sigma^*\}$$
$$= \Sigma^*\{ab\} = \Sigma^*\{a\}\{b\}$$
$$= (\{a\} \cup \{b\})^*\{a\}\{b\},$$

which we partially also presented in Example 2.1.20, shows that the language L_{ab} is rational.

b) It requires a short computation to check that we have

$$L_{\text{mod}} = \{w \in \Sigma^* \mid |w|_a \equiv 1 \,(\text{mod } 3)\}$$
$$= (\{b\}^*\{a\}\{b\}^*\{a\}\{b\}^*\{a\})^* \{b\}^*\{a\}\{b\}^*.$$

Using this identity, we obtain that the language L_{mod} is rational.

We already encountered the languages L_{ab} and L_{mod} in Example 2.2.9 when we showed their recognizability. We have now verified that they are rational as well. Indeed, Theorem 2.3.6 will exhibit that the notions of recognizability and rationality are equivalent for languages.

The following result comprises several closure properties of recognizable languages. These, being also of independent interest, will help us in the proof of Theorem 2.3.6.

2.3.4 Proposition.

a) The empty language is recognizable.

b) Every singleton language $\{a\}$ with $a \in \Sigma$ is recognizable.

c) Let L_1, L_2 and L be recognizable languages over Σ. Then the languages

$$\overline{L} = \Sigma^* \setminus L, \quad L^*, \quad L_1 L_2, \quad L_1 \cap L_2, \quad L_1 \cup L_2, \quad L_1 \setminus L_2$$

are again recognizable. In other words, the set of recognizable languages is closed under complement, star, concatenation, union, intersection as well as the relative complement.

d) Every finite language over Σ is recognizable.

Proof.

a) We consider the automaton $\mathcal{A} = (Q, I, T, F)$ with $Q = I = \{q\}$ and $T = F = \emptyset$ represented by the following state diagram:

Since $\mathcal{A}$ has no final state, there is in particular no successful path in $\mathcal{A}$. This results in $L(\mathcal{A}) = \emptyset$.

23

b) Given $w = a_1 \ldots a_n \in \Sigma^*$, the automaton

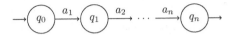

recognizes the singleton language $\{w\}$. In particular, the singleton language $\{a\}$ is recognizable for each $a \in \Sigma$.

c) Let $\mathcal{A} = (Q, I, T, F)$ and $\mathcal{A}_i = (Q_i, I_i, T_i, F_i)$ be automata such that $L(\mathcal{A}) = L$ and $L(\mathcal{A}_i) = L_i$ for $i = 1, 2$. Without loss of generality, we can assume that the automata $\mathcal{A}_1, \mathcal{A}_2$ and $\mathcal{A}$ are all complete and deterministic (see Remark 2.2.8). One can verify that the automaton $\overline{\mathcal{A}} = (Q, I, T, Q \setminus F)$ recognizes the complement $\overline{L} = \Sigma^* \setminus L$ (cf. Khoussainov and Nerode [17, Lemma 2.3.1]). Constructions of automata recognizing the concatenation $L_1 L_2$ and the star L^*, respectively, can be found in [17, Proof of Lemma 2.5.1]. It is straightforward to verify that the automaton $(Q_1 \times Q_2, I_1 \times I_2, T_\times, F_1 \times F_2)$ with

$$T_\times = \{((p_1, p_2), a, (q_1, q_2)) \mid (p_1, a, q_1) \in T_1 \text{ and } (p_2, a, q_2) \in T_2\}$$

recognizes the intersection $L_1 \cap L_2$. Intuitively, this automaton simultaneously simulates both $\mathcal{A}_1$ and $\mathcal{A}_2$. By the above, we conclude that also $L_1 \cup L_2 = \overline{(\overline{L_1} \cap \overline{L_2})}$ is recognizable. Alternatively, we can specify an explicit automaton recognizing the union $L_1 \cup L_2$. For this, we assume, without loss of generality, that the state sets Q_1 and Q_2 are disjoint (otherwise, replace Q_i by $Q_i \times \{i\}$ for $i = 1, 2$). Then it is straightforward to show that the automaton $(Q_1 \cup Q_2, I_1 \cup I_2, T_1 \cup T_2, F_1 \cup F_2)$ recognizes $L_1 \cup L_2$. Finally, the relative complement $L_1 \setminus L_2 = L_1 \cap \overline{L_2}$ is recognizable by the above.

d) By the proof of b), all singleton languages in Σ^+ are recognizable. Furthermore, the automaton

recognizes the singleton language $\{\varepsilon\}$. As every finite language can be written as a union of singleton languages, the claim follows directly by the above established closure properties.

$\square$

Proposition 2.3.4c) shows in particular that the rational operations union, concatenation and star preserve recognizability.

2.3.5 Example. Since any alphabet Σ is a finite language over itself, Proposition 2.3.4 implies that the language $(\Sigma\Sigma)^*$ containing precisely the words of even length is recognizable. In fact, an explicit complete deterministic automaton recognizing $(\Sigma\Sigma)^*$ is given by the following state diagram:

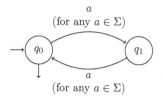

$$a$$
$$\text{(for any } a \in \Sigma\text{)}$$

$$a$$
$$\text{(for any } a \in \Sigma\text{)}$$

Furthermore, the language $(\Sigma\Sigma)^*$ is rational by Definition 2.3.1 and Remark 2.3.2, as the alphabet Σ can be regarded as a finite language over itself.

We are now ready to state the equivalence result of Kleene, which he first established in [18], and at least prove one direction.

2.3.6 Theorem (Kleene). *Let L be a language over Σ. Then L is rational if and only if L is recognizable.*

Proof. To prove that every rational language is recognizable, one proceeds by structural induction exploiting the closure properties of recognizable languages in Proposition 2.3.4. Detailed proofs of the backward direction can be found in Perrin [27, Theorem 4.1] and in Khoussainov and Nerode [17, § 2.5.3]. Their basic approach is to rewrite the language $L(\mathcal{A})$ recognized by some automaton $\mathcal{A}$ as a finite union of certain sublanguages and then to prove that the latter are rational. $\qquad\square$

Due to Theorem 2.3.6, there is no difference between the notions of recognizability and rationality for languages. However, more importantly, we now have two different but equivalent tools for the description of languages. Thus, whenever we want to prove a property of recognizable or rational languages, respectively, we can choose the more suitable representation.

2.3.7 Example. We have proved that the language $L_{\text{pump}} = \{a^n b^n \mid n \in \mathbb{N}\}$ is not recognizable in Example 2.2.13. We can conclude that the language

$$L = \{w \in \{a, b\}^* \mid |w|_a = |w|_b\}$$

is also not recognizable using Proposition 2.3.4. More precisely, we exploit that L_{pump} is the intersection of L and the recognizable language $\{a\}^*\{b\}^*$. Thus, if L were recognizable, then by Proposition 2.3.4c) the language L_{pump} would be recognizable as well, a contradiction. In particular, neither the language L nor L_{pump} is rational by Theorem 2.3.6.

Combining Proposition 2.3.4c) and Theorem 2.3.6, we obtain the following closure properties for rational languages.

2.3.8 Corollary. *Let L_1, L_2 and L be rational languages over Σ. Then the languages*

$$\overline{L} = \Sigma^* \setminus L, \quad L_1 \cap L_2, \quad L_1 \setminus L_2$$

are again rational. In other words, the set of rational languages is closed under complement, intersection and the relative complement.

Without using automata, it would certainly be very difficult to prove these closure properties.

In order to obtain another formalism for the representation of languages, we next employ logic – more precisely, monadic second-order logic – in the context of words and languages, respectively.

2.4. Monadic Second-Order Logic for Words

The study of the connections between Automata Theory and Formal Logic is as old as Theoretical Computer Science itself. Research on the logical aspects of the theory of finite automata began in the early 1960s with the work of Büchi [4], Elgot [15] and Trakhtenbrot [43] on monadic second-order logic in the context of words (cf. Straubing [41, page ix]). The basic idea is to use formulas of monadic second-order logic over a suitable signature in order to describe properties of words. In general, the approach of employing logics in Theoretical Computer Science plays an important role in the study of computational complexity, in particular in Descriptive Complexity Theory. These aspects are further developed in Straubing [41].

Since we are only interested in interpreting the formulas of monadic second-order logic in the context of words, we present a rather specialized formalism that is sufficient for this purpose. In this section, we assume the reader to be familiar with the fundamental notions from Mathematical Logic, and particularly from Model Theory. However, a detailed introduction of the general model-theoretic framework of monadic second-order logic can be found in Appendix A.1. In Appendix A.2 this general framework is carried out in the context of words, as applied in this section. We refer to these appendices whenever it is convenient.

In this section, we set up notation and terminology mainly following Droste [11, § 4].

We will soon give all the formal definitions of our framework, but first we present two introductory examples to illustrate the general concept.

2.4.1 Example. Consider the alphabet $\Sigma = \{a, b\}$ and the formula φ_1 given by

$$\exists x \, (P_a(x) \wedge \forall y \, x \leq y \wedge \forall z (x < z \to \neg P_a(z))).$$

How to read this formula? The letters x, y and z are individual variables interpreted as positions in a word, i.e. natural numbers in $\{1, \dots, n\}$ where n is the length of the respective word. The symbol P_a denotes a predicate, and e.g. $P_a(x)$ means "the xth letter of the word is a". The formula $\forall y \, x \leq y$ tells us that x is the first position of the word. In summary, the formula φ_1 says "the first letter of the word is a and there is no other a in the word". As b is the only letter in Σ besides a, the formula defines the rational language $\{a\}\{b\}^*$.

The next example is inspired by [41, Example II.1.c].

26

2.4.2 Example. Consider the formula φ_2 given by

$$\exists X (\forall x (\forall z\, x \leq z \rightarrow x \in X) \wedge$$
$$\forall x (\forall z\, z \leq x \rightarrow \neg x \in X) \wedge$$
$$\forall x\, \forall y\, (\mathrm{suc}(x,y) \rightarrow (x \in X \leftrightarrow \neg y \in X)))$$

How to read this formula? As above, the letters x, y and z are individual variables. The capital letter X is a set variable, which is interpreted as a set of positions in a word, i.e. a subset of $\{1, \ldots, n\}$ where n is the length of the respective word. Thus, the formula says "there exists a set X of positions such that ...". One can say that the variable X represents a predicate of positions in the word. As the predicate takes only a single argument, it is called a *monadic* predicate. The subformula

$$(\forall z\, x \leq z \rightarrow x \in X)$$

says that the set X contains the first position. Likewise, the subformula

$$(\forall z\, z \leq x \rightarrow \neg x \in X)$$

tells us that the last position does not belong to X. The expression $\mathrm{suc}(x,y)$ is an abbreviation for the formula

$$(x \leq y \wedge \neg y \leq x \wedge \forall z(z \leq x \vee y \leq z)).$$

Hence, the subformula $(\mathrm{suc}(x,y) \rightarrow (x \in X \leftrightarrow \neg y \in X))$ says that if x and y are two consecutive positions, then X contains exactly one of them. Thus, given some word w over Σ with $|w| = n$, we have to distinguish two cases to determine whether it fulfills the formula φ_2. If n is odd, then there does not exist a suitable set $X \subseteq \{1, \ldots, n\}$ of positions. In contrast, if n is even, then the set $X = \{1, 3, \ldots, n-1\}$ fulfills the scope of φ_2, since we have

$$1 \in X,\, 2 \notin X,\, 3 \in X,\, 4 \notin X,\, \ldots,\, n-1 \in X,\, n \notin X.$$

From this we can derive that the formula φ_2 defines the language $(\Sigma\Sigma)^*$ containing the words of even length, which we already encountered in Example 2.3.5. Definition 2.4.9 will exhibit that, indeed, also the empty word satisfies φ_2, since every formula of the form $\exists X \forall x\, \psi$ is satisfied by the empty word.

To formally introduce our logical framework, we have to specify its non-logical symbols. We consider the relational signature $\sigma_\Sigma = (\leq, (P_a)_{a \in \Sigma})$ where $\leq$ denotes a binary relation symbol and P_a a unary relation symbol for each letter $a \in \Sigma$. Moreover, we fix two countably infinite disjoint sets Var_1 and Var_2 of **first-order** and **second-order variables**, respectively, and we set $\mathrm{Var} := \mathrm{Var}_1 \cup \mathrm{Var}_2$. Occasionally, we refer to first-order variables as *individual variables* and to second-order variables as *set variables*.

We can now define the formal syntax of the logical statements in our framework.

2.4.3 Definition. We define the set of **monadic second-order formulas** over Σ, or the set of MSO(Σ)–formulas, inductively as follows:

(1) $P_a(x)$ is an MSO(Σ)–formula for each $a \in \Sigma$ and $x \in \mathrm{Var}_1$.

(2) $x \leq y$ is an MSO(Σ)–formula for any $x, y \in \mathrm{Var}_1$.

(3) $x \in X$ is an MSO(Σ)–formula for each $x \in \mathrm{Var}_1$ and $X \in \mathrm{Var}_2$.

(4) If φ is an MSO(Σ)–formula, then so is its **negation** $\neg\varphi$.

(5) If φ and ψ are MSO(Σ)–formulas, then so is their **disjunction** $(\varphi \vee \psi)$.

(6) If φ is an MSO(Σ)–formula and $x \in \mathrm{Var}_1$, then the **first-order existential quantification** $\exists x\, \varphi$ is again an MSO(Σ)–formula.

(7) If φ is an MSO(Σ)–formula and $X \in \mathrm{Var}_2$, then the **second-order existential quantification** $\exists X\varphi$ is again an MSO(Σ)–formula.

We write $\varphi \in$ MSO(Σ) if φ is an MSO(Σ)–formula. Moreover, we call the formulas obtained by steps (1)–(3) **atomic**.

We use standard expressions and abbreviations such as $x = y$ for $(x \leq y \wedge y \leq x)$, $(\varphi \wedge \psi)$ for $\neg(\neg\varphi \vee \neg\psi)$, and $\forall X\varphi$ for $\neg\exists X\neg\varphi$. Further expressions and abbreviations can be found in Appendix A.1 and Appendix A.2.

In order to interpret MSO(Σ)–formulas in words over Σ, we need to represent them as σ_Σ–structures.

2.4.4 Definition. Let w be a word over Σ.

a) We denote by
$$\mathrm{dom}(w) := \{1, \ldots, |w|\} \subseteq \mathbb{N}$$
the set of **positions** in w if w is non-empty. For the empty word we set $\mathrm{dom}(\varepsilon) := \emptyset$.

b) Given $i \in \mathrm{dom}(w)$, we denote by $w(i)$ the letter of Σ at position i in w, i.e. we have $w = w(1) \ldots w(n)$ with $n = |w|$. In effect, we identify the word w with the map $\mathrm{dom}(w) \to \Sigma$, $i \mapsto w(i)$.

c) We interpret $\leq$ as the usual linear ordering on $\mathrm{dom}(w)$ induced by the one on the natural numbers, i.e. $1 \leq 2 \leq \ldots \leq |w|$.

d) For $a \in \Sigma$ we set
$$P_a^w := \{i \in \mathrm{dom}(w) \mid w(i) = a\}.$$
For convenience, we simply write P_a for P_a^w if no confusion is likely to arise.

e) We identify the word w with the σ_Σ–structure
$$\underline{w} := (\mathrm{dom}(w), \leq, (P_a^w)_{a \in \Sigma}).$$

We observe that the sets $(P_a^w)_{a \in \Sigma}$ form a partition of $\mathrm{dom}(w)$. Moreover, the above described identification of words with σ_Σ–structures is justified since

$$w_1 = w_2 \iff \underline{w_1} = \underline{w_2}$$

holds for all words $w_1, w_2 \in \Sigma^*$. Indeed, the equality $\underline{w_1} = \underline{w_2}$ implies $|w_1| = |w_2|$ and $P_a^{w_1} = P_a^{w_2}$ for any letter $a \in \Sigma$. Thus, for each $i \in \mathrm{dom}(w_1) = \mathrm{dom}(w_2)$ and any letter $a \in \Sigma$ we obtain the equivalence $i \in P_a^{w_1} \iff i \in P_a^{w_2}$ and hence $w_1(i) = w_2(i)$. It follows $w_1 = w_2$. Similarly, one can prove the other implication. Further, we note that the empty word corresponds to the empty σ_Σ–structure $\underline{\varepsilon} = (\emptyset, \emptyset, (\emptyset)_{a \in \Sigma})$.

2.4.5 Example. Consider the alphabet $\Sigma = \{a, b, c, d\}$ and the word $w = abbacb$. Then we have

$$\mathrm{dom}(w) = \{1, 2, 3, 4, 5, 6\},$$
$$P_a = \{1, 4\},$$
$$P_b = \{2, 3, 6\},$$
$$P_c = \{5\},$$
$$P_d = \emptyset.$$

The introductory examples at the beginning of this section already give an intuition for the interpretation of MSO(Σ)–formulas. To formally define the semantics of MSO(Σ), we first need to specify *free occurrences* of variables in a formula.

2.4.6 Definition. Given $\varphi \in \mathrm{MSO}(\Sigma)$, we define the set $\mathrm{Free}(\varphi)$ of **free variables** of the formula φ by structural induction as follows:

$$\mathrm{Free}(P_a(x)) := \{x\},$$
$$\mathrm{Free}(x \leq y) := \{x, y\},$$
$$\mathrm{Free}(x \in X) := \{x, X\},$$
$$\mathrm{Free}(\neg \varphi) := \mathrm{Free}(\varphi),$$
$$\mathrm{Free}(\varphi \vee \psi) := (\mathrm{Free}(\varphi) \cup \mathrm{Free}(\psi)),$$
$$\mathrm{Free}(\exists x \, \varphi) := \mathrm{Free}(\varphi) \setminus \{x\},$$
$$\mathrm{Free}(\exists X \varphi) := \mathrm{Free}(\varphi) \setminus \{X\}.$$

A formula $\varphi \in \mathrm{MSO}(\Sigma)$ with $\mathrm{Free}(\varphi) = \emptyset$ is called an MSO(Σ)–**sentence**.

2.4.7 Example. For the formula $\varphi := (P_a(x) \vee \exists x \, x \leq y \vee \forall z \, z \in X)$ we obtain

$$\mathrm{Free}(\varphi) = \underbrace{\mathrm{Free}(P_a(x))}_{=\{x\}} \cup \underbrace{\mathrm{Free}(\exists x \, x \leq y)}_{=\{y\}} \cup \underbrace{\mathrm{Free}(\forall z \, z \in X)}_{=\{X\}} = \{x, y, X\}.$$

Note that the variable x occurs various times. Although its occurrence in the subformula $\exists x \, x \leq y$ is *bound*, x is a free variable of the formula φ, since its occurrence in the subformula $P_a(x)$ is free. In general, the set $\mathrm{Free}(\varphi)$ consists precisely of those variables that possess at least one free occurrence in φ.

The next definition provides a means to interpret the free variables of a formula. More precisely, free individual and set variables are substituted by positions and sets of positions in a word, respectively.

2.4.8 Definition. Let w be a word over Σ and $\mathcal{V} \subseteq \text{Var}$ be a finite set of variables.

a) We set $\mathcal{V}_i := (\mathcal{V} \cap \text{Var}_i)$ for $i = 1, 2$. Thus, $\mathcal{V}_1$ contains all individual variables in $\mathcal{V}$ and $\mathcal{V}_2$ contains all set variables in $\mathcal{V}$.

b) A $(\mathcal{V}, w)$–**assignment** is a map $\sigma \colon \mathcal{V} \to \text{dom}(w) \cup \mathcal{P}(\text{dom}(w))$ such that

$$\sigma(\mathcal{V}_1) \subseteq \text{dom}(w) \quad \text{and}$$
$$\sigma(\mathcal{V}_2) \subseteq \mathcal{P}(\text{dom}(w)).$$

By this, a $(\mathcal{V}, w)$–assignment maps first-order variables in $\mathcal{V}$ to positions and second-order variables in $\mathcal{V}$ to sets of positions in w.

c) Given a $(\mathcal{V}, w)$–assignment σ, a position $i \in \text{dom}(w)$ and a first-order variable $x \in \text{Var}_1$, we let $\sigma[x \mapsto i]$ be the $(\mathcal{V} \cup \{x\}, w)$–assignment mapping x to i and acting like σ elsewhere, i.e. $\sigma[x \mapsto i]$ satisfies

$$\sigma[x \mapsto i](x) = i \quad \text{and}$$
$$\sigma[x \mapsto i]\big|_{\mathcal{V} \setminus \{x\}} = \sigma\big|_{\mathcal{V} \setminus \{x\}}.$$

Similarly, if $X \in \text{Var}_2$ is a second-order variable and $I \subseteq \text{dom}(w)$ a set of positions, then we define a $(\mathcal{V} \cup \{X\}, w)$–assignment $\sigma[X \mapsto I]$ by

$$\sigma[X \mapsto I](X) = I \quad \text{and}$$
$$\sigma[X \mapsto I]\big|_{\mathcal{V} \setminus \{X\}} = \sigma\big|_{\mathcal{V} \setminus \{X\}}.$$

We are now ready to define the semantics of $\text{MSO}(\Sigma)$.

2.4.9 Definition. Let φ be an $\text{MSO}(\Sigma)$–formula, $\mathcal{V} \subseteq \text{Var}$ a finite set of variables such that $\text{Free}(\varphi) \subseteq \mathcal{V}$, $w \in \Sigma^+$ a non-empty word and σ a $(\mathcal{V}, w)$–assignment. We define the **satisfaction relation** $(w, \sigma) \models \varphi$, read (w, σ) **models** or **satisfies** φ, by structural induction as follows:

$$(w, \sigma) \models P_a(x) \quad :\Leftrightarrow \quad \sigma(x) \in P_a^w \quad :\Leftrightarrow \quad w(\sigma(x)) = a,$$
$$(w, \sigma) \models x \leq y \quad :\Leftrightarrow \quad \sigma(x) \leq \sigma(y),$$
$$(w, \sigma) \models x \in X \quad :\Leftrightarrow \quad \sigma(x) \in \sigma(X),$$
$$(w, \sigma) \models \neg\varphi \quad :\Leftrightarrow \quad (w, \sigma) \not\models \varphi \quad :\Leftrightarrow \quad \text{not } (w, \sigma) \models \varphi,$$
$$(w, \sigma) \models (\varphi_1 \vee \varphi_2) \quad :\Leftrightarrow \quad (w, \sigma) \models \varphi_1 \text{ or } (w, \sigma) \models \varphi_2,$$
$$(w, \sigma) \models \exists x\, \varphi \quad :\Leftrightarrow \quad (w, \sigma[x \mapsto i]) \models \varphi \text{ for some } i \in \text{dom}(w),$$
$$(w, \sigma) \models \exists X \varphi \quad :\Leftrightarrow \quad (w, \sigma[X \mapsto I]) \models \varphi \text{ for some } I \subseteq \text{dom}(w).$$

The semantics for the empty word are slightly different. The empty word satisfies all atomic $\mathrm{MSO}(\Sigma)$–formulas, i.e. the relations

$$\varepsilon \models P_a(x),$$
$$\varepsilon \models x \leq y,$$
$$\varepsilon \models x \in X$$

hold unconditionally. On the contrary, first-order existential quantifications are never satisfied by the empty word, i.e. we unconditionally have

$$\varepsilon \not\models \exists x \, \varphi.$$

The satisfaction of negations and disjunctions is defined as usual:

$$\varepsilon \models \neg\varphi \; :\Leftrightarrow \; \varepsilon \not\models \varphi \; :\Leftrightarrow \; \text{not } \varepsilon \models \varphi,$$
$$\varepsilon \models (\varphi_1 \vee \varphi_2) \; :\Leftrightarrow \; \varepsilon \models \varphi_1 \text{ or } \varepsilon \models \varphi_2.$$

Further, we set

$$\varepsilon \models \exists X \varphi \; :\Leftrightarrow \; \varepsilon \models \varphi.$$

A justification why it makes sense to define the semantics for the empty word in this way can be found in Remark A.1.13b).

2.4.10 Lemma. *Let φ be an $\mathrm{MSO}(\Sigma)$–formula, $\mathcal{V} \subseteq \mathrm{Var}$ a finite set of variables such that $\mathrm{Free}(\varphi) \subseteq \mathcal{V}$, $w \in \Sigma^+$ a non-empty word and σ a $(\mathcal{V}, w)$–assignment. We have*

$$(w, \sigma) \models \varphi \; \Leftrightarrow \; (w, \sigma|_{\mathrm{Free}(\varphi)}) \models \varphi$$

i.e. the satisfaction of φ only depends on w and the restriction $\sigma|_{\mathrm{Free}(\varphi)}$ of σ to $\mathrm{Free}(\varphi)$.

Proof. This result can be verified by structural induction (see Lemma A.1.15 and Remark A.2.5). □

Hence, whether or not we have $(w, \sigma) \models \varphi$ only depends on the word w and the values of free variables in φ under the assignment σ. As sentences do not possess free variables, Lemma 2.4.10 shows in particular that the satisfaction of a sentence does not depend on any variable assignment. More precisely, if φ is an $\mathrm{MSO}(\Sigma)$–sentence, $\mathcal{V}, \mathcal{V}' \subseteq \mathrm{Var}$ are finite sets of variables, $w \in \Sigma^+$ is a non-empty word, σ is a $(\mathcal{V}, w)$–assignment and σ' is a $(\mathcal{V}', w)$–assignment, then we have

$$(w, \sigma) \models \varphi \; \Leftrightarrow \; (w, \sigma') \models \varphi.$$

Therefore, we simply set

$$w \models \varphi \; :\Leftrightarrow \; (w, \epsilon) \models \varphi$$

where ϵ denotes the empty $(\mathrm{Free}(\varphi), w)$–assignment, i.e. the empty assignment from $\mathrm{Free}(\varphi) = \emptyset$ into $\mathrm{dom}(w) \cup \mathcal{P}(\mathrm{dom}(w))$.

2.4.11 Definition. Let φ be an MSO(Σ)–sentence. Then we call the language

$$L(\varphi) := \{w \in \Sigma^* \mid w \models \varphi\}$$

containing all words over Σ that satisfy φ the **language defined by the sentence** φ.

2.4.12 Example.

a) Consider the alphabet $\Sigma = \{a, b\}$ and the formula φ_1 given by

$$\exists x \, (P_a(x) \wedge \forall y \, x \leq y \wedge \forall z(x < z \to \neg P_a(z))),$$

which we already encountered in Example 2.4.1. First, we note that φ_1 is an MSO(Σ)–sentence since Free(φ_1) = $\emptyset$. As we already indicated in Example 2.4.1, the language defined by φ_1 is in fact given by $L(\varphi_1) = \{a\}\{b\}^*$, i.e. we have

$$w \models \varphi \; \Leftrightarrow \; w = ab^n \text{ for some } n \in \mathbb{N}$$

for each $w \in \Sigma^*$.

b) The formula φ_2 from Example 2.4.2 given by

$$\exists X \, (\forall x \, (\forall z \, x \leq z \to x \in X) \wedge$$
$$\forall x \, (\forall z \, z \leq x \to \neg x \in X) \wedge$$
$$\forall x \, \forall y \, (\mathrm{suc}(x, y) \to (x \in X \leftrightarrow \neg y \in X)))$$

is also an MSO(Σ)–sentence. Indeed, it defines the language

$$L(\varphi_2) = (\Sigma\Sigma)^*$$

consisting of all words over Σ of even length.

We observe that φ_1 contains only first-order quantifications while φ_2 contains a second-order quantification. Nevertheless, both sentences define recognizable languages.

2.4.13 Example. Consider the alphabet $\Sigma = \{a, b\}$. An MSO(Σ)–sentence defining the language $L_{ab} = \{w \in \Sigma^* \mid w \text{ ends with } ab\}$ is given by

$$\varphi := \exists x \, \exists y \, (\mathrm{suc}(x, y) \wedge \forall z(z \neq y \to z \leq x) \wedge P_a(x) \wedge P_b(y)).$$

Indeed, the sentence φ says "there exist two consecutive positions x and y that are the last positions of the word and the xth letter is a and the yth letter is b". The language L_{ab} is not only definable by some MSO(Σ)–sentence but is also recognizable as well as rational, as we have shown in Example 2.2.9a) and Example 2.3.3a). In particular, the MSO(Σ)–sentence φ is expressively equivalent to the automata presented in Example 2.2.9a).

In Section 2.3.6 we have seen that the formalism provided by the rational operations union, concatenation and star is equivalent to the one of finite automata. In the next section, it is our aim to prove that also monadic second-order logic provides a tool for the representation of languages that has the same expressive power as finite automata.

2.5. The Büchi–Elgot–Trakhtenbrot Theorem

The main result of this section is the Büchi–Elgot–Trakhtenbrot Theorem, which shows the equivalence in expressive power between finite automata and monadic second-order logic. Thus, this result, which dates back to 1960, establishes a very early connection between Mathematical Logic and Theoretical Computer Science, and it is considered the first result of Descriptive Complexity Theory. The theorem was first exhibited by Büchi [4] in 1960, and it was established independently by Elgot [15] and Trakhtenbrot [43] in 1961. Therefore, it is referred to as Büchi–Elgot–Trakhtenbrot Theorem.

Before we turn to the statement and proof of the Büchi–Elgot–Trakhtenbrot Theorem, we first explain the motivation behind their investigations on monadic second-order logic in the context of Automata Theory. In their joint work [3] from 1958, Büchi and Elgot (together with Wright) considered applications of Logic to Automata Theory. Conversely, they also wanted to study possible applications of Automata Theory to Logic, in particular to decision problems in Mathematical Logic (cf. [4, page 66] and [15, page 22]). More precisely, their aim was to produce procedures for deciding the truth of sentences, in particular of logical statements about the natural numbers. Furthermore, from Büchi's point of view, formulas of monadic second-order logic "seem to be more convenient" than rational expressions for formalizing conditions on the behavior of automata (cf. [4, page 66]). In the introduction of [15], Elgot considers, among others, the following question: Given an automaton and a *design requirement*[4], is there an algorithm deciding whether the automaton does satisfy the design requirement or not? His aim was to produce such an algorithm, in order to answer his question in the affirmative, or else to show that such an algorithm does not exist. Thus, Elgot considered the decidability of the *Model Checking Problem*. In fact, Corollary 2.5.16, which we deduce from the proof of the Büchi–Elgot–Trakhtenbrot Theorem, exhibits that the Model Checking Problem is decidable. Hence, Elgot's question can be answered in the affirmative.

2.5.1 Theorem (Büchi, Elgot and Trakhtenbrot). *Let L be a language over Σ. Then L is recognizable if and only if L is definable by some* MSO(Σ)*-sentence.*

We follow Droste [11, § 4] for the proof of Theorem 2.5.1. Alternative proofs can be found in Khoussainov and Nerode [17, § 2.10], Straubing [41, § III.1], Libkin [22, § 7.4] and Thomas [42, § 3.1].

The following proposition shows the forward direction of Theorem 2.5.1.

2.5.2 Proposition. *If L is a recognizable language over Σ, then L is definable by some* MSO(Σ)*-sentence.*

Proof. Let $\mathcal{A} = (Q, I, T, F)$ be a finite automaton over Σ recognizing L, i.e. such that $L(\mathcal{A}) = L$. As the state set Q of $\mathcal{A}$ is non-empty and finite, we can enumerate it

---

[4]Design requirements are specifications that the automaton shall fulfill, and in our context they are expressed by monadic second-order sentences

and write it as $Q = \{q_0, \ldots, q_m\}$ for some $m \in \mathbb{N}$. The basic idea of the proof is to encode a successful run P in $\mathcal{A}$ with label w by suitable sets

$$X_0, \ldots, X_m \subseteq \mathrm{dom}(w),$$

where X_i contains those positions $j \in \mathrm{dom}(w)$ for which we reach state q_i after executing $w(j)$ in P. [5] The existence of a successful run P with label w will then be equivalent to the existence of these sets. More precisely, we consider the MSO(Σ)–sentence

$$\varphi := \exists X_0, \ldots, X_m \, (\mathrm{partition}(X_0, \ldots, X_m) \wedge \varphi_{\mathrm{initial}} \wedge \varphi_{\mathrm{trans}} \wedge \varphi_{\mathrm{final}})$$

where $\mathrm{partition}(X_0, \ldots, X_m)$ is an abbreviation for

$$\forall x \bigvee_{i=0}^{m} \left(x \in X_i \wedge \bigwedge_{\substack{j=0 \\ \text{with } j \neq i}}^{m} x \notin X_j \right)$$

and the subformulas $\varphi_{\mathrm{initial}}, \varphi_{\mathrm{trans}}$ and φ_{final} are defined as follows:

$$\varphi_{\mathrm{initial}} := \exists x \left(\forall y \, x \leq y \wedge \left(\bigvee_{i=0}^{m} \bigvee_{\substack{(q,a) \in I \times \Sigma \\ \text{with } (q,a,q_i) \in T}} (P_a(x) \wedge x \in X_i) \right) \right)$$

$$\varphi_{\mathrm{trans}} := \forall x \, \forall y \left(\mathrm{suc}(x,y) \rightarrow \left(\bigvee_{i,j=0}^{m} \bigvee_{\substack{a \in \Sigma \\ \text{with } (q_i,a,q_j) \in T}} (x \in X_i \wedge P_a(y) \wedge y \in X_j) \right) \right)$$

$$\varphi_{\mathrm{final}} := \exists x \left(\forall y \, y \leq x \wedge \left(\bigvee_{\substack{i=0 \\ \text{with } q_i \in F}}^{m} x \in X_i \right) \right)$$

Observe that all disjunctions are finite, as the sets $I \subseteq Q$ and Σ are finite. Further, note that we consider an empty disjunction as $\bot$, i.e. as false (see Notation A.1.19). Intuitively, these formulas have the following interpretations:

- $\mathrm{partition}(X_0, \ldots, X_m)$ says "the sets $X_0, \ldots, X_m$ partition the set of positions".

- $\varphi_{\mathrm{initial}}$ says "there exists a first position and a suitable transition reading the first letter and starting in an initial state".

- φ_{trans} says "for consecutive positions there exists a corresponding transition in the automaton".

- φ_{final} says "there exists a last position and at this position we reach a final state".

---

[5] For instance, consider an automaton with state set $Q = \{q_0, q_1, q_2, q_3\}$. For the path

$$P = q_0 \xrightarrow{w(1)} q_1 \xrightarrow{w(2)} q_2 \xrightarrow{w(3)} q_0 \xrightarrow{w(4)} q_2$$

we have $X_0 = \{3\}, X_1 = \{1\}, X_2 = \{2, 4\}$ and $X_3 = \emptyset$.

For each word $w \in \Sigma^*$ with $w \models \varphi$ it must hold $w \neq \varepsilon$ due to the first-order existential quantifications in φ_{initial} and φ_{final}. However, we obtain:

Claim. *For every non-empty word $w \in \Sigma^+$, we have the following equivalence:*

$$w \in L(\mathcal{A}) \iff w \models \varphi.$$

Proof of Claim. Given a successful path

$$P = p_0 \xrightarrow{w(1)} p_1 \to \cdots \to p_{n-1} \xrightarrow{w(n)} p_n$$

with $n = |w| > 0$ in $\mathcal{A}$, one can verify that the sets $X_0, \ldots, X_m \subseteq \text{dom}(w)$ defined by

$$X_i := \{j \in \text{dom}(w) \mid p_j = q_i\} \quad (\text{for } i \in \{0, \ldots, m\})$$

fulfill the scope of the second-order quantification in φ. Thus, $w \in L(\mathcal{A})$ implies $w \models \varphi$. For the converse, we assume that there exist sets $X_0, \ldots, X_m$ that fulfill the scope of the second-order quantification in φ. Then the subformula partition$(X_0, \ldots, X_m)$ guarantees that for each position $j \in \text{dom}(w)$ there exists a unique state $p_j := q_i$ with $i \in \{0, \ldots, m\}$ and $j \in X_i$. Furthermore, the subformula φ_{initial} ensures that there is at least one $k \in \{0, \ldots, m\}$ such that $q_k \in I$ and $(q_k, w(1), p_1) \in T$. We choose one specific k fulfilling these conditions and set $p_0 := q_k$. Then

$$P = p_0 \xrightarrow{w(1)} p_1 \to \cdots \to p_{n-1} \xrightarrow{w(n)} p_n$$

with $n = |w|$ constitutes a successful path in $\mathcal{A}$ with label w. Indeed, by definition we have $p_0 = q_k \in I$ and $(p_0, w(1), p_1) \in T$, the subformula φ_{trans} ensures that $(p_{j-1}, w(j), p_j) \in T$ for any $j \in \text{dom}(w)$, and the subformula φ_{final} implies $p_n \in F$. Hence, we have $w \in L(\mathcal{A})$ if $w \models \varphi$. This completes the proof of the claim. $\diamond$

We infer from the claim that $L(\varphi) = L(\mathcal{A}) \setminus \{\varepsilon\}$. This results in $L(\varphi) = L(\mathcal{A}) = L$ if the automaton $\mathcal{A}$ does not recognize the empty word, i.e. if the sets I and F are disjoint. On the other hand, if $\varepsilon \in L(\mathcal{A})$, then we define another MSO(Σ)–sentence by

$$\psi := (\varphi \vee \forall x \, x < x)$$

to achieve

$$L(\psi) = L(\varphi) \cup \underbrace{L(\forall x \, x < x)}_{=\{\varepsilon\}} = L(\mathcal{A}).$$

This completes the proof of the proposition. $\square$

To prove the backward direction of the equivalence in Theorem 2.5.1, we wish to proceed by structural induction. For this, we need to associate languages not only to sentences but also to formulas with free variables. To say whether or not a word w satisfies a formula φ, we need an interpretation of its free variables, i.e. a (Free(φ), w)–assignment σ. Accordingly, it would then be natural to say that the language defined by a formula is the set consisting of all pairs (w, σ) that satisfy it. However, at first glance, this set does not seem to be a language. However, it is possible to encode pairs of the form (w, σ) as words over a certain alphabet. According to Straubing [41, page 20], the idea for this encoding comes from Perrin and Pin [26]. A very detailed explanation why their approach is quite natural can be found in Droste [11, page 47 f.].

2.5.3 Definition. Let $\mathcal{V} \subseteq \mathrm{Var}$ be a finite set of variables.

a) We define an **extended alphabet** $\Sigma_{\mathcal{V}} := \Sigma \times \{0,1\}^{\mathcal{V}}$, where $\{0,1\}^{\mathcal{V}}$ denotes the set of functions from $\mathcal{V}$ into $\{0,1\}$, as usual. For $\mathcal{V} = \emptyset$, we let $\Sigma_{\emptyset} := \Sigma$. Usually, we simply write $\Sigma_{\mathcal{V}}^*$ instead of $(\Sigma_{\mathcal{V}})^*$.

b) Let $w \in \Sigma^*$ be a word of length $|w| = n$ and σ a $(\mathcal{V}, w)$–assignment. We encode the pair (w, σ) as a word v of length n over the extended alphabet $\Sigma_{\mathcal{V}}$ with

$$v(i) := (a_i, \beta_i) \quad (\text{for } i = 1, \ldots, n),$$

where $a_i = w(i)$ and the map $\beta_i \colon \mathcal{V} \to \{0,1\}$ is given by

$$\beta_i(x) = \begin{cases} 1 & i = \sigma(x) \\ 0 & \text{otherwise} \end{cases} \quad (\text{for } x \in \mathcal{V}_1),$$

$$\beta_i(X) = \begin{cases} 1 & i \in \sigma(X) \\ 0 & \text{otherwise} \end{cases} \quad (\text{for } X \in \mathcal{V}_2).$$

c) Given a word $v \in \Sigma_{\mathcal{V}}^*$ of length $|v| = n$, we write $v(i) = (a_i, \beta_i)$ where $a_i \in \Sigma$ and $\beta_i \in \{0,1\}^{\mathcal{V}}$ for $i = 1, \ldots, n$. We call v **valid** if either $v = \varepsilon$ or if for each first-order variable $x \in \mathcal{V}_1$ there exists a unique $i \in \mathrm{dom}(v)$ such that $\beta_i(x) = 1$ (and $\beta_j(x) = 0$ for any $j \neq i$).

d) We set $N_{\mathcal{V}} := \{v \in \Sigma_{\mathcal{V}}^* \mid v \text{ is valid}\}$.

e) Let $v \in N_{\mathcal{V}}$ be a non-empty valid word with $v(i) = (a_i, \beta_i) \in \Sigma_{\mathcal{V}}$ for $i = 1, \ldots, n$ where $n = |v|$. We decode v to obtain a pair (w, σ) consisting of a word w over Σ and a $(\mathcal{V}, w)$–assignment σ as follows: The word over Σ is given by $w = a_1 \ldots a_n$ and the $(\mathcal{V}, w)$–assignment σ maps each first-order variable $x \in \mathcal{V}_1$ to the unique position $i \in \mathrm{dom}(v) = \mathrm{dom}(w)$ with $\beta_i(x) = 1$ and each second-order variable $X \in \mathcal{V}_2$ to the set $\{i \in \mathrm{dom}(w) \mid \beta_i(X) = 1\}$. Decoding the empty word $\varepsilon \in N_{\mathcal{V}}$ yields the empty word $\varepsilon \in \Sigma^*$.

One can verify that the above described encoding and decoding provide a bijective correspondence between the empty word and pairs consisting of a word $w \in \Sigma^+$ and a $(\mathcal{V}, w)$–assignment σ on the one hand and valid words over the extended alphabet $\Sigma_{\mathcal{V}}$ on the other hand. For convenience, we often write $(w, \sigma) \in N_{\mathcal{V}}$ meaning that we obtain the pair (w, σ) by decoding a valid word $v \in \Sigma_{\mathcal{V}}^*$. Further, we observe:

$$N_{\mathcal{V}} = \Sigma_{\mathcal{V}}^* \iff \mathcal{V}_1 = \emptyset.$$

An alternative encoding can be found in Straubing [41, pages 14–17]. The main difference is that he works with the extended alphabet $\Sigma \times \mathcal{P}(\mathcal{V}_1) \times \mathcal{P}(\mathcal{V}_2)$ instead of the alphabet $\Sigma_{\mathcal{V}} = \Sigma \times \{0,1\}^{\mathcal{V}}$. Since for any set M there is a bijective correspondence between its power set $\mathcal{P}(M)$ and the set of functions $\{0,1\}^M$, the two approaches are equivalent.

2.5.4 Remark. Let $\mathcal{V} \subseteq \mathrm{Var}$ be a set consisting of n first-order variables and m second-order variables, where $n, m \in \mathbb{N}$. Then we can write $\mathcal{V}_1 = \{x_1, \ldots, x_n\}$ and $\mathcal{V}_2 = \{X_1, \ldots, X_m\}$. Usually, we represent a letter $(a, \beta) \in \Sigma_{\mathcal{V}}$ by the following column vector in $\Sigma \times \{0,1\}^n \times \{0,1\}^m$:

$$
\begin{pmatrix}
a \\
\beta(x_1) \\
\vdots \\
\beta(x_n) \\
\beta(X_1) \\
\vdots \\
\beta(X_m)
\end{pmatrix}
\begin{array}{l}
\\
\longleftarrow x_1\text{-row} \\
\vdots \\
\longleftarrow x_n\text{-row} \\
\longleftarrow X_1\text{-row} \\
\vdots \\
\longleftarrow X_m\text{-row}
\end{array}
$$

2.5.5 Example. Let $\Sigma = \{a, b, c\}$ and $\mathcal{V} = \{x, y, X\}$.

a) Consider the word $w = cbab$ and the $(\mathcal{V}, w)$–assignment σ given by

$$\sigma(x) = 2, \ \sigma(y) = 3, \ \sigma(X) = \{1, 3, 4\}.$$

The encoding of the pair (w, σ) is given by the following word over $\Sigma_{\mathcal{V}}$:

$$
v_1 = \begin{pmatrix} c \\ 0 \\ 0 \\ 1 \end{pmatrix}\begin{pmatrix} b \\ 1 \\ 0 \\ 0 \end{pmatrix}\begin{pmatrix} a \\ 0 \\ 1 \\ 1 \end{pmatrix}\begin{pmatrix} b \\ 0 \\ 0 \\ 1 \end{pmatrix}
\begin{array}{l} \\ \longleftarrow x\text{-row} \\ \longleftarrow y\text{-row} \\ \longleftarrow X\text{-row} \end{array}
$$

b) A word over $\Sigma_{\mathcal{V}}$ that is not valid is given by:

$$
v_2 = \begin{pmatrix} a \\ 0 \\ 0 \\ 1 \end{pmatrix}\begin{pmatrix} b \\ 1 \\ 0 \\ 0 \end{pmatrix}\begin{pmatrix} b \\ 1 \\ 1 \\ 1 \end{pmatrix}\begin{pmatrix} a \\ 0 \\ 0 \\ 1 \end{pmatrix}
\begin{array}{l} \\ \longleftarrow x\text{-row} \\ \longleftarrow y\text{-row} \\ \longleftarrow X\text{-row} \end{array}
$$

The reason for v_2 not being valid is that there are two 1's in the x–row. In particular, the word v_2 cannot be decoded to obtain a pair (w, σ).

c) Consider the following valid word over $\Sigma_{\mathcal{V}}$:

$$
v_3 = \begin{pmatrix} a \\ 1 \\ 0 \\ 1 \end{pmatrix}\begin{pmatrix} a \\ 0 \\ 0 \\ 1 \end{pmatrix}\begin{pmatrix} b \\ 0 \\ 0 \\ 1 \end{pmatrix}\begin{pmatrix} c \\ 0 \\ 0 \\ 0 \end{pmatrix}\begin{pmatrix} b \\ 0 \\ 1 \\ 0 \end{pmatrix}
\begin{array}{l} \\ \longleftarrow x\text{-row} \\ \longleftarrow y\text{-row} \\ \longleftarrow X\text{-row} \end{array}
$$

Decoding the word v_3 yields the pair (w, σ) consisting of the word $w = aabcb$ and the $(\mathcal{V}, w)$–assignment σ satisfying

$$\sigma(x) = 1, \; \sigma(y) = 5, \; \sigma(X) = \{1, 2, 3\}.$$

2.5.6 Definition. Let φ be an MSO(Σ)–formula and $\mathcal{V} \subseteq$ Var a finite set of variables such that Free(φ) $\subseteq \mathcal{V}$. We define

$$L_\mathcal{V}(\varphi) := \{(w, \sigma) \in N_\mathcal{V} \mid (w, \sigma) \models \varphi\}.$$

Moreover, we set $\Sigma_\varphi := \Sigma_{\text{Free}(\varphi)}$, $N_\varphi := N_{\text{Free}(\varphi)}$ and $L(\varphi) := L_{\text{Free}(\varphi)}(\varphi)$. Then $L(\varphi)$ is called the **language defined by the formula** φ.

This definition is consistent with our previous notation of $L(\varphi)$ for an MSO(Σ)–sentence φ from Definition 2.4.11. Indeed, if Free(φ) $= \emptyset$, then we have $\Sigma_\varphi = \Sigma_\emptyset = \Sigma$ by definition. Thus, we obtain

$$L_\emptyset(\varphi) = L(\varphi) = \{w \in \Sigma^* \mid w \models \varphi\}.$$

In Lemma 2.4.10 we have seen that the satisfaction of a formula in a word only depends on the interpretation of its free variables. The following result provides a corresponding statement with respect to the recognizability of the language defined by a formula. A detailed proof can be found in Droste [11, Lemma 4.8].

2.5.7 Lemma. *Let φ be an MSO(Σ)–formula and $\mathcal{V} \subseteq$ Var a finite set of variables such that Free(φ) $\subseteq \mathcal{V}$. Then $L_\mathcal{V}(\varphi)$ is recognizable if and only if $L(\varphi)$ is recognizable.*

As mentioned above, our goal is to show by structural induction that $L(\varphi)$ is a recognizable language over Σ_φ for any MSO(Σ)–formula φ. In the induction step, the following result will be used.

2.5.8 Lemma. *Let $\mathcal{V} \subseteq$ Var be a finite set of variables. The language $N_\mathcal{V}$ consisting of all valid words over the extended alphabet $\Sigma_\mathcal{V}$ is recognizable.*

Proof. We show that $\overline{N_\mathcal{V}} = \Sigma_\mathcal{V}^* \setminus N_\mathcal{V}$ is recognizable. Then the claim follows by applying Proposition 2.3.4c). For each $x \in \mathcal{V}_1$ and $k = 0, 1$ we put

$$\Sigma_\mathcal{V}^{x,k} := \{(a, \beta) \in \Sigma_\mathcal{V} \mid \beta(x) = k\}.$$

Note that $\Sigma_\mathcal{V}^{x,k} \subseteq \Sigma_\mathcal{V}$ both are finite languages. We now have

$$\overline{N_\mathcal{V}} = \bigcup_{x \in \mathcal{V}_1} \left((\Sigma_\mathcal{V}^{x,0})^+ \mathbin{\dot{\cup}} (\Sigma_\mathcal{V}^* \cdot \Sigma_\mathcal{V}^{x,1} \cdot \Sigma_\mathcal{V}^* \cdot \Sigma_\mathcal{V}^{x,1} \cdot \Sigma_\mathcal{V}^*) \right).$$

Indeed, all words in this union are non-empty, words in $(\Sigma_\mathcal{V}^{x,0})^+$ have no 1 in the x–row, and words in the language-theoretic concatenation on the right-hand side of the disjoint union have more than one 1 in the x–row. Since the set $\mathcal{V}_1$ is finite, the language $\overline{N_\mathcal{V}}$ is recognizable by Proposition 2.3.4. $\square$

2.5.9 Lemma. *Let φ be an atomic $\mathrm{MSO}(\Sigma)$-formula. Then $L(\varphi)$ is recognizable.*

Proof. Let v be an arbitrary non-empty valid word over the extended alphabet Σ_φ, i.e. $\varepsilon \neq v \in N_\varphi$, and write $v(i) = (a_i, \beta_i) \in \Sigma_\varphi$ for $i = 1, \ldots, |v|$. As usual, decoding the valid word v yields a pair (w, σ) consisting of a word w over Σ and a $(\mathrm{Free}(\varphi), w)$–assignment σ. We now have to distinguish three cases:

(1) If $\varphi = P_a(x)$, then we have $\mathrm{Free}(\varphi) = \{x\}$ and we obtain

$$
\begin{aligned}
v \in L(\varphi) &\Leftrightarrow (w, \sigma) \models P_a(x) \\
&\Leftrightarrow \sigma(x) \in P_a \\
&\Leftrightarrow w(i) = a \text{ for the unique } i \in \mathrm{dom}(w) \\
&\quad \text{with } \beta_i(x) = 1.
\end{aligned}
$$

Further, we have $\varepsilon \in L(\varphi)$ since by definition $\varepsilon \models P_a(x)$. Thus, the language $L(\varphi)$ is recognized by the following automaton:

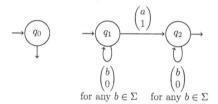

(2) If $\varphi = (x \leq y)$, then we have $\mathrm{Free}(\varphi) = \{x, y\}$ and we obtain

$$
\begin{aligned}
v \in L(\varphi) &\Leftrightarrow (w, \sigma) \models x \leq y \\
&\Leftrightarrow \sigma(x) \leq \sigma(y) \\
&\Leftrightarrow i \leq j \text{ for the unique } i, j \in \mathrm{dom}(w) \\
&\quad \text{with } \beta_i(x) = 1 = \beta_j(y).
\end{aligned}
$$

Further, we have $\varepsilon \in L(\varphi)$ since by definition $\varepsilon \models x \leq y$. Thus, the language $L(\varphi)$ is recognized by the following automaton:

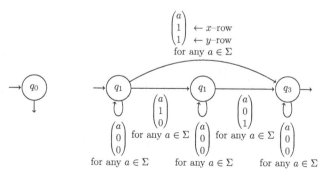

(3) If $\varphi = (x \in X)$, then we have $\text{Free}(\varphi) = \{x, X\}$ and we obtain

$$
\begin{aligned}
v \in L(\varphi) \;\; &\Leftrightarrow \;\; (w, \sigma) \models x \in X \\
&\Leftrightarrow \;\; \sigma(x) \in \sigma(X) \\
&\Leftrightarrow \;\; \beta_i(X) = 1 \text{ for the unique } i \in \text{dom}(w) \\
&\quad\;\; \text{with } \beta_i(x) = 1.
\end{aligned}
$$

Further, we have $\varepsilon \in L(\varphi)$ since by definition $\varepsilon \models x \in X$. Thus, the language $L(\varphi)$ is recognized by the following automaton:

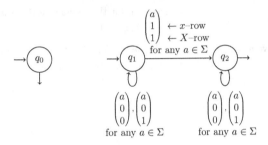

$\square$

In order to prove the backward direction of the equivalence in Theorem 2.5.1, we proceed by structural induction and show the stronger fact that $L(\varphi)$ is a recognizable language for any MSO(Σ)–formula φ. Having already considered the case of atomic MSO(Σ)–formulas in Lemma 2.5.9, we now turn to negation and disjunction. More precisely, we show that both negation and disjunction preserve recognizability in the following sense:

2.5.10 Lemma. *Let φ and ψ be MSO(Σ)–formulas.*

a) If $L(\varphi)$ and $L(\psi)$ are recognizable, then $L(\varphi \vee \psi)$ is again recognizable.

b) If $L(\varphi)$ is recognizable, then $L(\neg\varphi)$ is again recognizable.

Proof.

a) First, we observe that in general we do not have $L(\varphi \vee \psi) = L(\varphi) \cup L(\psi)$ since possibly $\text{Free}(\varphi) \neq \text{Free}(\psi)$. Thus, $L(\varphi), L(\psi)$ and $L(\varphi \vee \psi)$ may be languages over different alphabets. Therefore, we put $\mathcal{V} = L(\varphi \vee \psi) = \text{Free}(\varphi) \cup \text{Free}(\psi)$. By assumption and Lemma 2.5.7, $L_\mathcal{V}(\varphi)$ and $L_\mathcal{V}(\psi)$ are recognizable. Hence, their union

$$
L_\mathcal{V}(\varphi \vee \psi) = L_\mathcal{V}(\varphi) \cup L_\mathcal{V}(\psi)
$$

is recognizable by Proposition 2.3.4c). Lemma 2.5.7 then implies that $L(\varphi \vee \psi)$ is recognizable.

b) We first recall that $\mathrm{Free}(\neg\varphi) = \mathrm{Free}(\varphi)$ and thus $N_{\neg\varphi} = N_\varphi$. Then we have

$$
\begin{aligned}
L(\neg\varphi) &= \{(w,\sigma) \in N_\varphi \mid (w,\sigma) \models \neg\varphi\} \\
&= \{(w,\sigma) \in N_\varphi \mid (w,\sigma) \not\models \varphi\} \\
&= N_\varphi \cap \overline{L(\varphi)}.
\end{aligned}
$$

By Proposition 2.3.4c) the complement $\overline{L(\varphi)} = (\Sigma_\mathcal{V})^* \setminus L(\varphi)$ is recognizable. Hence, the claim follows by Lemma 2.5.8 and Proposition 2.3.4c).

$\square$

Next, we exhibit that also existential quantifications preserve recognizability.

2.5.11 Lemma. *Let φ be an $\mathrm{MSO}(\Sigma)$–formula such that $L(\varphi)$ is recognizable. Then the languages $L(\exists x\,\varphi)$ and $L(\exists X\varphi)$ are recognizable as well.*

Proof. First, we consider the language $L(\exists X\varphi)$ defined by second-order existential quantification. We put $\mathcal{V} = \mathrm{Free}(\exists X\varphi)$, i.e. $X \notin \mathcal{V}$, and let $\pi \colon \Sigma^*_{\mathcal{V}\cup\{X\}} \to \Sigma^*_\mathcal{V}$ be the unique length-preserving monoid homomorphism extending the map

$$
\Sigma_{\mathcal{V}\cup\{X\}} \to \Sigma_\mathcal{V}, \; (a,\beta) \mapsto (a,\beta\big|_\mathcal{V})
$$

(see Lemma 2.1.12 and Remark 2.1.15). Thus, π deletes the X–row in the vector representation of letters in $\Sigma_{\mathcal{V}\cup\{X\}}$ (see Remark 2.5.4), and therefore we understand it as a projection. As X is a second-order variable, it is straightforward to verify the equalities

$$
\pi(N_{\mathcal{V}\cup\{X\}}) = N_\mathcal{V} \quad \text{and} \quad \pi^{-1}(N_\mathcal{V}) = N_{\mathcal{V}\cup\{X\}},
$$

i.e. the projection π preserves validness. Hence, for each non-empty $(w,\sigma) \in \Sigma^+_\mathcal{V}$ we obtain

$$
\begin{aligned}
(w,\sigma) \in L(\exists X\varphi) \;&\Leftrightarrow\; (w,\sigma) \in N_\mathcal{V} \text{ and } (w,\sigma) \models \exists X\varphi \\
&\Leftrightarrow\; (w,\sigma) \in N_\mathcal{V} \text{ and there exists } I \subseteq \mathrm{dom}(w) \\
&\qquad \text{such that } (w,\sigma[X \mapsto I]) \models \varphi \\
&\Leftrightarrow\; \text{there exists } I \subseteq \mathrm{dom}(w) \text{ such that} \\
&\qquad (w,\sigma[X \mapsto I]) \in L_{\mathcal{V}\cup\{X\}}(\varphi) \\
&\Leftrightarrow\; (w,\sigma) \in \pi\big(L_{\mathcal{V}\cup\{X\}}(\varphi)\big)
\end{aligned}
$$

since $\pi\big(w,\sigma[X \mapsto I]\big) = (w,\sigma)$ for $I \subseteq \mathrm{dom}(w)$. Further, we have $\varepsilon \in L(\exists X\varphi)$ if and only if $\varepsilon \in L(\varphi)$ by our definition of the semantics for the empty word. Consequently, we obtain

$$
L(\exists X\varphi) = \pi\big(L_{\mathcal{V}\cup\{X\}}(\varphi)\big),
$$

which is recognizable by the assumption, Lemma 2.5.7 and Proposition 2.2.11.

Next, we turn to the language $L(\exists x\,\varphi)$ defined by first-order existential quantification. As before, we put $\mathcal{V} = \text{Free}(\exists x\,\varphi)$, i.e. $x \notin \mathcal{V}$, and let $\pi\colon \Sigma_{\mathcal{V}\cup\{x\}}^* \to \Sigma_{\mathcal{V}}^*$ be the unique length-preserving monoid homomorphism extending the map

$$\Sigma_{\mathcal{V}\cup\{x\}} \to \Sigma_{\mathcal{V}}, \ (a,\beta) \mapsto (a,\beta|_{\mathcal{V}})$$

(see Lemma 2.1.12 and Remark 2.1.15). Thus, π deletes the x–row in the vector representation of letters in $\Sigma_{\mathcal{V}\cup\{x\}}$ (see Remark 2.5.4), and therefore we understand it as a projection. Note that we still have the equality

$$\pi(N_{\mathcal{V}\cup\{x\}}) = N_{\mathcal{V}} \ \text{ but } \ N_{\mathcal{V}\cup\{x\}} \subsetneq \pi^{-1}(N_{\mathcal{V}}),$$

as the validness of words over $\Sigma_{\mathcal{V}\cup\{x\}}^*$ depends on the evaluation at the first-order variable x which is deleted by π. However, for each non-empty $(w,\sigma) \in \Sigma_{\mathcal{V}}^+$ we obtain

$$
\begin{aligned}
(w,\sigma) \in L(\exists x\,\varphi) \ &\Leftrightarrow\ (w,\sigma) \in N_{\mathcal{V}} \text{ and } (w,\sigma) \models \exists x\,\varphi \\
&\Leftrightarrow\ (w,\sigma) \in N_{\mathcal{V}} \text{ and there exists } i \in \text{dom}(w) \\
&\quad\ \text{such that } (w,\sigma[x \mapsto i]) \models \varphi \\
&\Leftrightarrow\ \text{there exists } i \in \text{dom}(w) \text{ such that} \\
&\quad\ (w,\sigma[x \mapsto i]) \in L_{\mathcal{V}\cup\{x\}}(\varphi) \\
&\Leftrightarrow\ (w,\sigma) \in \pi\big(L_{\mathcal{V}\cup\{x\}}(\varphi)\big)
\end{aligned}
$$

since $\pi\,(w,\sigma[x \mapsto i]) = (w,\sigma)$ for any $i \in \text{dom}(w)$, and we have $(w,\sigma[x \mapsto i]) \in N_{\mathcal{V}\cup\{x\}}$ if and only if $(w,\sigma) \in N_{\mathcal{V}}$. We further have $\varepsilon \notin L(\exists x\,\varphi)$ as by definition $\varepsilon \not\models \exists x\,\varphi$. Consequently, we obtain

$$L(\exists x\,\varphi) = \pi\big(L_{\mathcal{V}\cup\{x\}}(\varphi)\big) \setminus \{\varepsilon\},$$

which is recognizable by the assumption, Lemma 2.5.7, Proposition 2.2.11 and Proposition 2.3.4. $\qquad\square$

2.5.12 Proposition. *Let φ be an MSO(Σ)–formula and $\mathcal{V} \subseteq \text{Var}$ a finite set of variables such that $\text{Free}(\varphi) \subseteq \mathcal{V}$. Then $L(\varphi)$ and $L_{\mathcal{V}}(\varphi)$ are recognizable.*

Proof. First, we proceed by structural induction to show that $L(\varphi)$ is a recognizable language over Σ_φ. If φ is an atomic MSO(Σ)–formula then $L(\varphi)$ is recognizable by Lemma 2.5.9. The induction step follows directly from Lemma 2.5.10 and Lemma 2.5.11. Hence, $L(\varphi)$ is recognizable for each MSO(Σ)–formula. The recognizability of the language $L_{\mathcal{V}}(\varphi)$ then follows directly from Lemma 2.5.7. $\qquad\square$

Proof of Theorem 2.5.1. Proposition 2.5.2 implies that every recognizable language is definable by some MSO(Σ)–sentence. For the converse, let φ be some MSO(Σ)–sentence. Then $L(\varphi) = L_{\mathcal{V}}(\varphi)$ with $\mathcal{V} = \text{Free}(\varphi) = \emptyset$ is recognizable by Proposition 2.5.12. Hence, every language that is definable by some MSO(Σ)–sentence is recognizable. This completes the proof of the theorem. $\qquad\square$

An immediate consequence of the proof of Theorem 2.5.1 is that existential $MSO(\Sigma)$–sentences have the same expressive power as $MSO(\Sigma)$–sentences in general, as we explain in the following.

2.5.13 Definition. Let φ be an $MSO(\Sigma)$–formula. We call φ **existential** if it is of the form $\exists X_0, \ldots, X_n \, \psi$ for some $n \in \mathbb{N}$ and an $MSO(\Sigma)$–formula ψ containing no second-order quantifications.

2.5.14 Corollary. *Let L be a language over Σ. Then L is recognizable if and only if L is definable by some existential $MSO(\Sigma)$–sentence. In particular, every $MSO(\Sigma)$–sentence is equivalent to an existential $MSO(\Sigma)$–sentence in the sense of defining the same language.*

Proof. The proof of Proposition 2.5.2 shows that for each recognizable language $L \subseteq \Sigma^*$ there is an $MSO(\Sigma)$–sentence ψ such that $L(\psi) = L$. More precisely, depending on whether or not the language L contains the empty word, this sentence is either of the form

$$\exists X_0 \ldots X_m \, \xi$$

or of the form

$$\exists X_0 \ldots X_m \, \xi \vee \forall x \, x < x$$

for some $m \in \mathbb{N}$ and an $MSO(\Sigma)$–formula ξ that contains no second-order quantifications. If $\varepsilon \notin L$, then the $MSO(\Sigma)$–sentence $\psi = \exists X_0 \ldots X_m \, \xi$ is itself existential, and if $\varepsilon \in L$, then the $MSO(\Sigma)$–sentence

$$\psi' = \exists X_0 \ldots X_m \, (\xi \vee \forall x \, x < x)$$

is existential and fulfills $L(\psi') = L(\psi) = L$. Hence, every recognizable language is definable by some existential $MSO(\Sigma)$–sentence. Applying Theorem 2.5.1, this shows in particular that for each $MSO(\Sigma)$–sentence φ there is an existential $MSO(\Sigma)$–sentence ψ such that $L(\psi) = L(\varphi)$, i.e. ψ and φ are equivalent in the sense of defining the same language. The backward direction of the first statement follows directly from Proposition 2.5.12. □

2.5.15 Example. Recall that $L_{\mathrm{mod}} = \{w \in \Sigma^* \mid |w|_a \equiv 1 \,(\mathrm{mod}\,3)\}$ is a recognizable language over $\Sigma = \{a, b\}$ (see Example 2.2.9b)). Corollary 2.5.14 implies that L_{mod} is definable by some existential $MSO(\Sigma)$–sentence. More precisely, following the proof of Proposition 2.5.2 one can construct an explicit existential $MSO(\Sigma)$–sentence φ such that $L(\varphi) = L_{\mathrm{mod}}$, i.e.

$$w \models \varphi \Leftrightarrow |w|_a \equiv 1 \,(\mathrm{mod}\,3)$$

for any $w \in \Sigma^*$.

Due to Theorem 2.5.1, the recognizable languages are precisely those definable in monadic second-order logic for words. Thus, we can use $MSO(\Sigma)$–sentences in order to describe properties of finite automata and recognizable languages, respectively. Furthermore, we observe that the proofs of both implications in Theorem 2.5.1 are constructive. More precisely, they provide effective procedures or algorithms P, P' such that:

- Given an automaton $\mathcal{A}$ over Σ, algorithm P computes an MSO(Σ)–sentence $\varphi_{\mathcal{A}}$ that satisfies $L(\varphi_{\mathcal{A}}) = L(\mathcal{A})$.

- Given an MSO(Σ)–sentence φ, algorithm P' computes an automaton $\mathcal{A}_{\varphi}$ over Σ that satisfies $L(\mathcal{A}_{\varphi}) = L(\varphi)$.

As a consequence, we obtain the following decidability properties of monadic second-order logic or particularly of MSO(Σ)–sentences, respectively.

2.5.16 Corollary. *Let φ and φ' be MSO(Σ)–sentences. Then the following problems are decidable:*

a) Given a word w over Σ, does w satisfy φ?
(Word Problem)

b) Is there a word that satisfies φ?
(Satisfiability Problem)

c) Is φ satisfied by all words over Σ?
(Validity Problem)

d) Is φ satisfied by infinitely many words?
(Infinity Problem)

e) Is the language $L(\varphi)$ contained in the language $L(\varphi')$?
(Inclusion Problem)

f) Do φ and φ' define the same language?
(Equivalence Problem)

g) Given a finite automaton $\mathcal{A}$, is the language $L(\mathcal{A})$ contained in $L(\varphi)$?
(Model Checking Problem)

Proof. Following the proof of Proposition 2.5.12 we can effectively construct finite automata $\mathcal{A}$ and $\mathcal{A}'$ such that $L(\mathcal{A}) = L(\varphi)$ and $L(\mathcal{A}') = L(\varphi')$. Indeed, all results that we used in the proofs of Lemma 2.5.10 and Lemma 2.5.11 come with effective procedures. Moreover, a word w over Σ satisfies an MSO(Σ)–sentence φ if and only if w belongs to the language $L(\varphi)$ defined by φ. We note further that the Satisfiability Problem and the Validity Problem correspond to the Emptiness Problem and the Universality Problem in the context of finite automata. Thus, we can apply Theorem 2.2.14 to obtain that all problems are decidable. $\square$

As we have already mentioned at the beginning of this section, these decidability results were one of the main intentions of Büchi and Elgot's investigations on the connections between finite automata and monadic second-order logic. The decidability of the Satisfiability Problem is of independent interest, as it implies that there exists an algorithm for deciding whether a given MSO(Σ)–sentence is *satisfiable*, i.e. whether or not there exists a word that satisfies it. Similarly, the decidability of the Validity Problem means that there is a procedure determining for a given MSO(Σ)–sentence whether it is *valid*, i.e. whether or not it is satisfied by all words over Σ. The Satisfiability Problem is *dual* to Validity Problem in the following sense: A given MSO(Σ)–sentence φ is valid if and only if its negation $\neg\varphi$ is unsatisfiable, i.e. we have

$$L(\varphi) = \Sigma^* \ \Leftrightarrow \ L(\neg\varphi) = \emptyset.$$

2.5.17 Example. We want to explain the purpose of model checking. The general idea is that one is given a model of a system (e.g. a circuit or a protocol) and is confronted with the question whether this model meets a certain formal *specification* or a certain *design requirement* (e.g. a safety or liveness property). In our context, the system is modeled by an automaton $\mathcal{A}$ and the specification is given by a sentence φ of monadic second-order logic. Thus, we want to check whether all words w that are recognized by $\mathcal{A}$ satisfy φ, i.e. whether we have

$$w \in L(\mathcal{A}) \ \Rightarrow \ w \models \varphi$$

for any word w. If this is not the case, then there must be words $w \in L(\mathcal{A})$ that violate the specification φ, which then can be analyzed further. For instance, let us again consider the example of a snack machine (see Example 2.2.5). The letters in the alphabet $\Sigma = \{\texttt{pay}, \texttt{cancel}, \texttt{choose}, \texttt{remove}\}$ represent actions and thus words can be understood as action sequences or processes. Now consider the MSO(Σ)–sentence φ given by

$$\forall x \, \exists y \, ((\mathrm{suc}(x,y) \wedge P_{\texttt{choose}}(x)) \rightarrow P_{\texttt{remove}}(y)).$$

Roughly speaking, φ says that every customer that has chosen a snack does not go away empty-handed. More precisely, an action sequence fulfills φ if the action $\texttt{choose}$ is directly followed by the action $\texttt{remove}$, i.e. the machine gives out the chosen snack. Clearly, every snack machine should meet this requirement (at least from the point of view of a customer). By Corollary 2.5.16 the Model Checking Problem is decidable. In particular, there exists a program that checks whether our model of a snack machine from Example 2.2.5 meets the specification given by φ.

For first-order logic, denoted FO(Σ), there are also characterizations similar to those in Theorem 2.3.6 and Theorem 2.5.1. More precisely, combining Straubing [41, Theorem VI.1.1] and Droste [11, Proposition 3.3 and Theorem 3.5] we obtain:

2.5.18 Theorem (Schützenberger, McNaughton and Papert). *For any language L over Σ the following conditions are equivalent:*

(i) The language L is recognizable and aperiodic (cf. [11, Definition 3.2]).

(ii) The language L is star-free (cf. [11, Definition 3.1]).

(iii) The language L is definable by some FO(Σ)–sentence (see Definition A.2.2).

Originally, the equivalence of (i) and (ii) was pointed out by Schützenberger [39] in 1965, and the equivalence of (ii) and (iii) was first established by McNaughton and Papert [25] in 1971. As a direct consequence of Theorem 2.5.18, the language $(\Sigma\Sigma)^*$

containing the words of even length, which is definable by some $\mathrm{MSO}(\Sigma)$–sentence (see Example 2.4.2), is not definable by some $\mathrm{FO}(\Sigma)$–sentence (cf. [11, Corollary 4.14]). This shows in particular that monadic second-order logic is more expressive than first-order logic.

Let us return to monadic second-order logic. To show that the language $L(\varphi)$ defined by some $\mathrm{MSO}(\Sigma)$–sentence is recognizable we have proceeded by structural induction and thus have proved an even stronger statement for formulas in general (see Proposition 2.5.12). By similarly generalizing the other implication in Theorem 2.5.1, we obtain the following statement:

2.5.19 Theorem. *Let* $\mathcal{V} \subseteq \mathrm{Var}$ *be a finite set of variables. Given an $\mathrm{MSO}(\Sigma)$–formula φ with $\mathrm{Free}(\varphi) \subseteq \mathcal{V}$, the language $L_\mathcal{V}(\varphi)$ over the extended alphabet $\Sigma_\mathcal{V}$ is recognizable. Viceversa, if L is a recognizable language over $\Sigma_\mathcal{V}$, then there exists an $\mathrm{MSO}(\Sigma)$–formula φ with $\mathrm{Free}(\varphi) \subseteq \mathcal{V}$ such that*

$$L \cap N_\mathcal{V} = L_\mathcal{V}(\varphi). \tag{$*$}$$

The rest of this section is dedicated to the proof of this result. First, we observe that it is a natural generalization of Theorem 2.5.1. More precisely, the special case $\mathcal{V} = \emptyset$ in Theorem 2.5.19 corresponds precisely to the statement of Theorem 2.5.1. Indeed, an $\mathrm{MSO}(\Sigma)$–formula φ with $\mathrm{Free}(\varphi) \subseteq \emptyset$ is an $\mathrm{MSO}(\Sigma)$–sentence, and we have

$$N_\emptyset = \Sigma_\emptyset = \Sigma \text{ as well as } L_\emptyset(\varphi) = L(\varphi).$$

Hence, the main difference between the two theorems is that Theorem 2.5.1 just deals with sentences, whereas Theorem 2.5.19 deals with formulas in general. While multiple proofs for Theorem 2.5.1 have been published, a proof for Theorem 2.5.19 seems not to be available in the literature. In the rest of this section, we wish to fill this gap.

Let us have a closer look at the second part of Theorem 2.5.19. Thus, we assume that we are given a recognizable language L over the extended alphabet $\Sigma_\mathcal{V}$. Then Theorem 2.5.1 ensures the existence of an $\mathrm{MSO}(\Sigma_\mathcal{V})$–sentence ψ such that

$$L = L(\psi). \tag{$**$}$$

If we compare the formula φ from Theorem 2.5.19 with the sentence ψ, then we observe a major difference: While φ is a monadic second-order formula over the alphabet Σ, the sentence ψ is a monadic second-order formula over the extended alphabet $\Sigma_\mathcal{V}$. Thus, the underlying logical signatures differ. More precisely, ψ may contain atomic subformulas of the form $P_{(a,\beta)}(x)$ where $(a,\beta) \in \Sigma_\mathcal{V}$ is a letter in the extended alphabet. However, we want to continue working in $\mathrm{MSO}(\Sigma)$ and not switch to $\mathrm{MSO}(\Sigma_\mathcal{V})$. Hence, our idea is to replace any atomic $\mathrm{MSO}(\Sigma_\mathcal{V})$–formula in ψ of the form $P_{(a,\beta)}(x)$ by a suitable $\mathrm{MSO}(\Sigma)$–formula in order to obtain an equivalent $\mathrm{MSO}(\Sigma)$–formula φ with $\mathrm{Free}(\varphi) \subseteq \mathcal{V}$ in the following sense:

$$v \models \psi \iff (w,\sigma) \models \varphi$$

where $v \in N_\mathcal{V}$, and (w,σ) is the pair obtained by decoding the valid word v.

From this it follows

$$L(\psi) \cap N_{\mathcal{V}} = L_{\mathcal{V}}(\varphi),$$

which together with (∗∗) implies (∗). In summary, our core idea for the proof of Theorem 2.5.19 is to apply Theorem 2.5.1 and to perform certain syntactical substitutions. Before we pass to the details of the proof, we ultimately note that the intersection with the language $N_{\mathcal{V}}$ in (∗) is necessary, since a recognizable language $L \subseteq \Sigma_{\mathcal{V}}$ may contain words that are not valid. For instance, the language $L \subseteq \Sigma_{\{x\}}$ recognized by the automaton

$\leftarrow$ x-row

contains the non-valid word $\begin{pmatrix} a \\ 1 \end{pmatrix} \begin{pmatrix} a \\ 1 \end{pmatrix} \notin N_{\{x\}}$.

2.5.20 Definition. Let $\mathcal{V} \subseteq \mathrm{Var}$ be a finite set of variables, $(a, \beta) \in \Sigma_{\mathcal{V}}$ a letter in the extended alphabet, and $x \in \mathrm{Var}_1 \setminus \mathcal{V}$ a first-order variable not contained in $\mathcal{V}$. We define the $\mathrm{MSO}(\Sigma)$–formula $\varphi_{(a,\beta)}(x)$ by the disjunction

$$\varphi_{(a,\beta)}(x) := \left(\varphi^+_{(a,\beta)}(x) \vee \forall z\, z < z \right)$$

where the subformula $\varphi^+_{(a,\beta)}(x) \in \mathrm{MSO}(\Sigma)$ is given by the conjunction

$$\left(P_a(x) \wedge \bigwedge_{\substack{y \in \mathcal{V}_1 \\ \beta(y)=1}} x = y \wedge \bigwedge_{\substack{y \in \mathcal{V}_1 \\ \beta(y)=0}} x \neq y \wedge \bigwedge_{\substack{Y \in \mathcal{V}_2 \\ \beta(Y)=1}} x \in Y \wedge \bigwedge_{\substack{Y \in \mathcal{V}_2 \\ \beta(Y)=0}} x \notin Y \right).$$

Note that Free $\left(\varphi_{(a,\beta)}(x) \right) = $ Free $\left(\varphi^+_{(a,\beta)}(x) \right) = \mathcal{V} \,\dot\cup\, \{x\}$.

2.5.21 Remark. Let v be a valid word over $\Sigma_{\mathcal{V}}$, i.e. $v \in N_{\mathcal{V}}$. In order to determine whether the decoding of v satisfies the formulas $\varphi_{(a,\beta)}$ and $\varphi^+_{(a,\beta)}$, respectively, we distinguish two cases:

(1) If v is non-empty, then decoding v yields a pair (w, σ) consisting of a non-empty word w over Σ and a $(\mathcal{V}, w)$–assignment σ. Now we clearly have

$$(w, \sigma) \not\models \forall z\, z < z.$$

Thus, in this case the satisfaction of $\varphi_{(a,\beta)}(x)$ only depends on the satisfaction of its subformula $\varphi^+_{(a,\beta)}(x)$, i.e. we have

$$(w, \sigma) \models \varphi_{(a,\beta)}(x) \iff (w, \sigma) \models \varphi^+_{(a,\beta)}(x).$$

(2) If v is the empty word, then it is possibly the case that

$$\varepsilon \not\models \varphi^+_{(a,\beta)}(x).$$

For instance, if we have $\beta(Y) = 0$ for some second-order variable $Y \in \mathcal{V}_2$, then the conjunction contains $x \notin Y$ as a subformula and this yields $\varepsilon \not\models \varphi^+_{(a,\beta)}(x)$ as by definition $\varepsilon \models x \in Y$. However, we achieve

$$\varepsilon \models \varphi_{(a,\beta)}(x),$$

since by definition we have $\varepsilon \models \forall z \, z < z$.

The purpose of $\varphi_{(a,\beta)}(x)$ is to replace the atomic $\mathrm{MSO}(\Sigma_{\mathcal{V}})$–formula $P_{(a,\beta)}(x)$. Indeed, the two are equivalent in the following sense:

2.5.22 Lemma. *Let $\mathcal{V} \subseteq \mathrm{Var}$ be a finite set of variables, $(a,\beta) \in \Sigma_{\mathcal{V}}$ a letter in the extended alphabet, $x \in \mathrm{Var}_1 \setminus \mathcal{V}$ a first-order variable not contained in $\mathcal{V}$, and $v \in \Sigma^*_{\mathcal{V}}$ a valid word, i.e. $v \in N_{\mathcal{V}}$.*

(1) If v is non-empty, then we decode it to obtain a pair (w, σ) consisting of a non-empty word w over Σ and a $(\mathcal{V}, w)$–assignment σ, as usual. In this case we obtain

$$(v, \tau) \models P_{(a,\beta)}(x) \quad (in \ \mathrm{MSO}(\Sigma_{\mathcal{V}}))$$
$$\Leftrightarrow \ (w, \sigma[x \mapsto \tau(x)]) \models \varphi_{(a,\beta)}(x) \quad (in \ \mathrm{MSO}(\Sigma))$$

for each $(\{x\}, v)$–assignment τ.

(2) Decoding the empty word $v = \varepsilon \in N_{\mathcal{V}}$ yields the empty word $\varepsilon \in \Sigma^*$ and we obtain that both

$$\varepsilon \models P_{(a,\beta)}(x) \quad (in \ \mathrm{MSO}(\Sigma_{\mathcal{V}}))$$
$$and \ \ \varepsilon \models \varphi_{(a,\beta)}(x) \quad (in \ \mathrm{MSO}(\Sigma))$$

hold unconditionally.

Proof. First, we assume that v is non-empty and we denote by (w, σ) the pair obtained by decoding v. We recall that $\mathrm{dom}(w) = \mathrm{dom}(v)$. Further, we let $\tau \colon \{x\} \mapsto \mathrm{dom}(v)$ be a $(\{x\}, v)$–assignment. Now by the definition of the semantics we have

$$(v, \tau) \models P_{(a,\beta)}(x) \ \Leftrightarrow \ v(\tau(x)) = (a, \beta).$$

If we decode the right-hand side, then we obtain the conditions

- $w(\tau(x)) = a$,

- $\tau(x) = \sigma(y)$ for each $y \in \mathcal{V}_1$ with $\beta(y) = 1$,

- $\tau(x) \neq \sigma(y)$ for each $y \in \mathcal{V}_1$ with $\beta(y) = 0$,

- $\tau(x) \in \sigma(y)$ for each $Y \in \mathcal{V}_2$ with $\beta(Y) = 1$,

- $\tau(x) \notin \sigma(y)$ for each $Y \in \mathcal{V}_2$ with $\beta(Y) = 0$.

Since we assume that x is not contained in $\mathcal{V}$, the restriction $\sigma[x \mapsto \tau(x)]\big|_{\mathcal{V}}$ coincides with σ, and thus the above conditions are equivalent to

$$(w, \sigma[x \mapsto \tau(x)]) \models \varphi^+_{(a,\beta)}(x).$$

This equivalence together with our considerations in Remark 2.5.21(1) proves the first part of the lemma. The second part regarding the empty word follows directly from the definition of the semantics and our considerations in Remark 2.5.21(2). $\quad\square$

Exploiting this equivalence for atomic formulas, every $\mathrm{MSO}(\Sigma_{\mathcal{V}})$–formula can be replaced by an equivalent $\mathrm{MSO}(\Sigma)$–formula in the following sense:

2.5.23 Proposition. *Let $\mathcal{V} \subseteq \mathrm{Var}$ be a finite set of variables and ψ an $\mathrm{MSO}(\Sigma_{\mathcal{V}})$–formula such that no variable that occurs in ψ is contained in $\mathcal{V}$. Further, let φ be the $\mathrm{MSO}(\Sigma)$–formula which we obtain by replacing each atomic subformula in ψ of the form $P_{(a,\beta)}(x)$ by the $\mathrm{MSO}(\Sigma)$–formula $\varphi_{(a,\beta)}(x)$.*

a) The set $\mathrm{Free}(\varphi)$ of free variables in φ is contained in the disjoint union $\mathcal{V} \,\dot\cup\, \mathrm{Free}(\psi)$.

b) Let $v \in \Sigma^*_{\mathcal{V}}$ be a non-empty valid word, i.e. $\varepsilon \neq v \in N_{\mathcal{V}}$, (w, σ) the pair obtained by decoding v, and τ a $(\mathrm{Free}(\psi), v)$–assignment. Then we have

$$(v, \tau) \models \psi \quad (in\ \mathrm{MSO}(\Sigma_{\mathcal{V}}))$$
$$\Leftrightarrow \quad (w, \sigma[\tau]) \models \varphi \quad (in\ \mathrm{MSO}(\Sigma))$$

where $\sigma[\tau]$ denotes the $(\mathcal{V} \,\dot\cup\, \mathrm{Free}(\psi), w)$–assignment defined by

$$\sigma[\tau]\big|_{\mathcal{V}} = \sigma \ and\ \sigma[\tau]\big|_{\mathrm{Free}(\psi)} = \tau.$$

Decoding the empty word $v = \varepsilon \in N_{\mathcal{V}}$ yields the empty word $\varepsilon \in \Sigma^*$ and we obtain

$$\varepsilon \models \psi \quad (in\ \mathrm{MSO}(\Sigma_{\mathcal{V}}))$$
$$\Leftrightarrow \quad \varepsilon \models \varphi \quad (in\ \mathrm{MSO}(\Sigma))$$

Proof. During the proof, let $v \in N_{\mathcal{V}}$ be non-empty and (w, σ) denote the pair obtained by decoding v. We proceed by induction to simultaneously prove a) and b). Thus, we distinguish the following cases:

- If ψ is atomic and of the form $P_{(a,\beta)}(x)$, then $\varphi = \varphi_{(a,\beta)}(x)$ fulfills claim a) since $\mathrm{Free}(\varphi) = \mathcal{V} \cup \{x\} = \mathcal{V} \cup \mathrm{Free}(\psi)$. By assumption we further have $x \notin \mathcal{V}$ and thus b) follows directly from Lemma 2.5.22, since the $(\mathcal{V} \,\dot\cup\, \{x\})$–assignment $\sigma[\tau]$ coincides with $\sigma[x \mapsto \tau(x)]$.

- If ψ is atomic and of the form $x \leq y$, then $\varphi = \psi$ clearly fulfills claim a) as $\mathrm{Free}(\varphi) = \mathrm{Free}(\psi)$. Furthermore, we obtain

$$(v, \tau) \models x \leq y \Leftrightarrow \tau(x) \leq \tau(y)$$
$$\Leftrightarrow \sigma[\tau](x) \leq \sigma[\tau](y)$$
$$\Leftrightarrow (w, \sigma[\tau]) \models x \leq y$$

since the restriction $\sigma[\tau]|_{\mathrm{Free}(\psi)}$ coincides with τ.

Regarding the empty word, the satisfaction $\varepsilon \models x \leq y$ holds unconditionally both in $\mathrm{MSO}(\Sigma_\mathcal{V})$ and in $\mathrm{MSO}(\Sigma)$ by definition. Moreover, one proceeds analogously to prove a) and b) if the atomic $\mathrm{MSO}(\Sigma_\mathcal{V})$–formula ψ has the form $x \in X$.

- If ψ is a disjunction $\psi_1 \vee \psi_2$ of $\mathrm{MSO}(\Sigma_\mathcal{V})$–formulas ψ_1, ψ_2 fulfilling a) and b), then the corresponding $\mathrm{MSO}(\Sigma)$–formula φ is given by the disjunction $\varphi_1 \vee \varphi_2$ of the $\mathrm{MSO}(\Sigma)$–formulas corresponding to ψ_1 and ψ_2, respectively. Thus, we obtain $\mathrm{Free}(\varphi) = \mathrm{Free}(\varphi_1) \cup \mathrm{Free}(\varphi_2) \subseteq \mathcal{V} \cup \mathrm{Free}(\psi_1) \cup \mathrm{Free}(\psi_2) = \mathcal{V} \cup \mathrm{Free}(\psi)$, which proves a). For $i = 1, 2$ and each $(\mathrm{Free}(\psi), v)$–assignment τ we achieve by assumption and Lemma 2.4.10 the equivalences

$$(v, \tau) \models \psi_i \Leftrightarrow (v, \tau|_{\mathrm{Free}(\psi_i)}) \models \psi_i$$
$$\Leftrightarrow (w, \sigma[\tau|_{\mathrm{Free}(\psi_i)}]) \models \varphi_i$$
$$\Leftrightarrow (w, \sigma[\tau]) \models \varphi_i.$$

As a consequence, we obtain

$$(v, \tau) \models \psi_1 \vee \psi_2 \Leftrightarrow (v, \tau) \models \psi_1 \text{ or } (v, \tau) \models \psi_2$$
$$\Leftrightarrow (w, \sigma[\tau]) \models \varphi_1 \text{ or } (w, \sigma[\tau]) \models \varphi_2$$
$$\Leftrightarrow (w, \sigma[\tau]) \models \varphi_1 \vee \varphi_2.$$

This establishes the first part of b) regarding non-empty valid words. Finally, the verification of the equivalence regarding the empty word is straightforward. Moreover, one proceeds analogously if ψ is a negation $\neg \psi'$ of an $\mathrm{MSO}(\Sigma_\mathcal{V})$–formula ψ' fulfilling a) and b).

- If ψ is a first-order existential quantification $\exists x\, \psi'$ with an $\mathrm{MSO}(\Sigma_\mathcal{V})$–formula ψ' fulfilling a) and b), then the corresponding $\mathrm{MSO}(\Sigma)$–formula φ is given by $\exists x\, \varphi'$ where φ' is the $\mathrm{MSO}(\Sigma)$–formula corresponding to ψ'. Since by assumption we have $x \notin \mathcal{V}$, we obtain

$$\mathrm{Free}(\varphi) = \mathrm{Free}(\varphi') \setminus \{x\}$$
$$\subseteq (\mathcal{V} \cup \mathrm{Free}(\psi')) \setminus \{x\} = \mathcal{V} \cup (\mathrm{Free}(\psi') \setminus \{x\}) = \mathcal{V} \cup \mathrm{Free}(\psi),$$

which proves a). Furthermore, for each $(\mathrm{Free}(\psi), v)$–assignment τ and each position $i \in \mathrm{dom}(v) = \mathrm{dom}(w)$ we have $\sigma[(\tau[x \mapsto i])] = (\sigma[\tau])[x \mapsto i]$.

Hence, applying Lemma 2.4.10 and the assumption on ψ' yields

$$(v, \tau) \models \exists x\, \psi' \Leftrightarrow (v, \tau[x \mapsto i]) \models \psi' \text{ for some } i \in \text{dom}(v)$$
$$\Leftrightarrow (w, \sigma[(\tau[x \mapsto i])]) \models \varphi' \text{ for some } i \in \text{dom}(v) = \text{dom}(w)$$
$$\Leftrightarrow (w, (\sigma[\tau])[x \mapsto i]) \models \varphi' \text{ for some } i \in \text{dom}(w)$$
$$\Leftrightarrow (w, \sigma[\tau]) \models \exists x\, \varphi'.$$

For the empty word we both have $\varepsilon \not\models \exists x\, \psi'$ and $\varepsilon \not\models \exists x\, \varphi'$. If ψ is a second-order existential quantification $\exists X\, \psi'$ with an $\text{MSO}(\Sigma_\mathcal{V})$–formula ψ' fulfilling a) and b), then one proceeds analogously. However, there is a minor difference regarding the empty word: In the case of second-order existential quantifications we obtain

$$\varepsilon \models \exists X\, \psi' \Leftrightarrow \varepsilon \models \psi' \Leftrightarrow \varepsilon \models \varphi' \Leftrightarrow \varepsilon \models \exists X\, \varphi'.$$

This completes the proof of the proposition. $\qquad\qquad\square$

2.5.24 Example. Consider again the set $\mathcal{V} = \{x\}$ and the automaton $\mathcal{A}$ over the extended alphabet $\Sigma_\mathcal{V}$ given by the state diagram

$\leftarrow x\text{-row}$

which we already encountered above. Recall from Remark 2.5.4 that the column vector represents the letter $(a, \beta) \in \Sigma_\mathcal{V} = \Sigma_{\{x\}}$ where the assignment $\beta \colon \{x\} \to \{0,1\}$ is defined by $\beta(x) = 1$. The automaton $\mathcal{A}$ recognizes the language

$$L = \left\{ \begin{pmatrix} a \\ 1 \end{pmatrix}^n \,\middle|\, n \in \mathbb{N} \right\} \subseteq \Sigma_\mathcal{V}^*.$$

By Theorem 2.5.1 there is some $\text{MSO}(\Sigma_\mathcal{V})$–sentence ψ such that $L(\psi) = L$. Indeed, e.g. the $\text{MSO}(\Sigma_\mathcal{V})$–sentence

$$\psi = \forall y\, P_{(a,\beta)}(y)$$

fulfills $L(\psi) = L$. If we replace the atomic subformula $P_{(a,\beta)}(y)$ in ψ by the $\text{MSO}(\Sigma)$–formula $\varphi_{(a,\beta)}(y)$, then we obtain the $\text{MSO}(\Sigma)$–formula

$$\varphi = \forall y\, ((P_a(y) \wedge y = x) \vee \forall z\, z < z)$$

with $\text{Free}(\varphi) = \{x\} = \mathcal{V}$. With our considerations from Remark 2.5.21 it is easy to verify that the simpler $\text{MSO}(\Sigma)$–formula

$$\varphi' = \forall y\, (P_a(y) \wedge y = x)$$

defines the same language as φ. More precisely, both formulas define the language

$$L' = \left\{ \varepsilon, \begin{pmatrix} a \\ 1 \end{pmatrix} \right\} = \left\{ \begin{pmatrix} a \\ 1 \end{pmatrix}^0, \begin{pmatrix} a \\ 1 \end{pmatrix}^1 \right\} = L \cap N_\mathcal{V}.$$

In fact, we have $\varepsilon \models \varphi'$, since φ' is a first-order universal quantification. Furthermore, for each non-empty valid word $v \in \Sigma_\mathcal{V}^*$, i.e. $\varepsilon \neq v \in N_\mathcal{V}$, and its decoding (w, σ), consisting of a non-empty word $w \in \Sigma^+$ and a $(\{x\}, w)$-assignment σ, we obtain

$$(w, \sigma) \models \forall y \, (P_a(y) \wedge y = x)$$
$$\Leftrightarrow (w, \sigma[y \mapsto i]) \models (P_a(y) \wedge y = x) \text{ for all } i \in \mathrm{dom}(w)$$
$$\Leftrightarrow w(i) = a \text{ and } i = \sigma(x) \text{ for all } i \in \mathrm{dom}(w)$$
$$\Leftrightarrow w = a \text{ and } \sigma(x) = 1 \text{ (i.e. } |w| = 1)$$
$$\Leftrightarrow v = \begin{pmatrix} a \\ 1 \end{pmatrix}.$$

Thus, we achieve $L(\varphi') = L_\mathcal{V}(\varphi') = L'$. Consequently, the formulas φ and φ' fulfill equation $(*)$, respectively.

We are now ready to give a detailed proof of Theorem 2.5.19. For the second part, we basically proceed as in Example 2.5.24.

Proof of Theorem 2.5.19. By Proposition 2.5.12, $L_\mathcal{V}(\varphi)$ is a recognizable language over the extended alphabet $\Sigma_\mathcal{V}$ for any $\mathrm{MSO}(\Sigma)$–formula φ with $\mathrm{Free}(\varphi) \subseteq \mathcal{V}$. For the converse, let L be a recognizable language over $\Sigma_\mathcal{V}$. Then by Theorem 2.5.1 there exists an $\mathrm{MSO}(\Sigma_\mathcal{V})$–sentence such that $L = L(\psi)$. Without loss of generality we may assume that the variables occurring in ψ are not contained in $\mathcal{V}$, since we can rename them otherwise. By replacing each atomic $\mathrm{MSO}(\Sigma_\mathcal{V})$–formula in ψ of the form $P_{(a,\beta)}(x)$ by the $\mathrm{MSO}(\Sigma)$–formula $\varphi_{(a,\beta)}(x)$, we obtain an $\mathrm{MSO}(\Sigma)$–formula φ. By Proposition 2.5.23a) this formula fulfills

$$\mathrm{Free}(\varphi) \subseteq \underbrace{\mathrm{Free}(\psi)}_{=\emptyset} \,\dot\cup\, \mathcal{V} = \mathcal{V}.$$

Now let $v \in \Sigma_\mathcal{V}^*$ be a non-empty valid word, i.e. $\varepsilon \neq v \in N_\mathcal{V}$, and denote by (w, σ) the pair obtained by decoding v. Then Proposition 2.5.23b) guarantees that

$$(v, \epsilon) \models \psi \quad (\text{in } \mathrm{MSO}(\Sigma_\mathcal{V}))$$
$$\Leftrightarrow (w, \sigma[\epsilon]) \models \varphi \quad (\text{in } \mathrm{MSO}(\Sigma))$$

where ϵ denotes the empty $(\emptyset, v)$-assignment. Consequently, we obtain

$$v \in L(\psi) = L$$
$$\Leftrightarrow (w, \sigma) \in L_\mathcal{V}(\varphi)$$

since $\sigma[\epsilon]$ coincides with σ. As a final observation, also by Proposition 2.5.23b) the empty word fulfills

$$\varepsilon \models \psi \quad (\text{in } \mathrm{MSO}(\Sigma_\mathcal{V}))$$
$$\Leftrightarrow \varepsilon \models \varphi \quad (\text{in } \mathrm{MSO}(\Sigma)),$$

i.e. we have $\varepsilon \in L(\psi) = L$ if and only if $\varepsilon \in L_\mathcal{V}(\varphi)$, which ultimately establishes the equation $(*)$. $\qquad\square$

Recall that every recognizable language over an alphabet Γ is definable by some *existential* MSO(Γ)–sentence (see Corollary 2.5.14). In particular, the formula ψ in the proof of Theorem 2.5.19 can be chosen to be existential. As we only replace atomic MSO($\Sigma_{\mathcal{V}}$)–subformulas in ψ by suitable MSO(Σ)–formulas that contain no second-order quantifications in the proof of Theorem 2.5.19, the substitute φ of ψ is again existential. Hence, we obtain a similar result as Corollary 2.5.14 for formulas in general.

2.5.25 Corollary. *Let $\mathcal{V} \subseteq$ Var be a finite set of variables and L a recognizable language over the extended alphabet $\Sigma_{\mathcal{V}}$. Then there is an existential MSO(Σ)–formula φ with Free(φ) $\subseteq \mathcal{V}$ such that*

$$L \cap N_{\mathcal{V}} = L_{\mathcal{V}}(\varphi).$$

In particular, for each MSO(Σ)–formula there is some equivalent existential MSO(Σ)–formula in the sense of defining the same language.

Before we move on to the next chapter, we give a graphical overview of the main results we have exhibited so far:

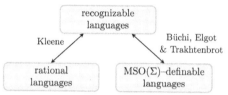

There are many other characterizations of recognizable languages, and in particular, there are further formalisms for the modeling of languages. However, we restrict ourselves to the ones provided by Theorem 2.3.6 and Theorem 2.5.1, as the main objective of this work is to extend these language-theoretic results to the realm of formal power series in the following chapters. More precisely, our investigations will ultimately result in the following diagram:

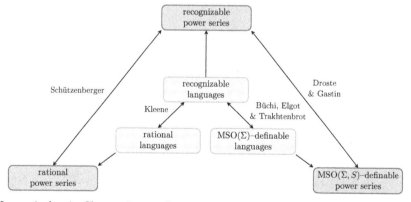

In particular, in Chapter 5 we will state and prove further generalizations of Theorem 2.5.1 based on an approach entirely different to that of Theorem 2.5.19.

3. Weighted Automata

In this chapter, we start our investigation of weighted automata. Weighted automata can be viewed as a quantitative extension of classical automata. More precisely, a weighted automaton is a finite automaton whose transitions and states are equipped with weights. These weights may model e.g. the cost involved when executing a transition, the amount of resources or time needed for this, or the probability or reliability of its successful execution (cf. Droste, Kuich and Vogler [8, page v]). While classical automata recognize words or languages, respectively, a weighted automaton recognizes a map that associates to each word the total weight of its execution. Such maps are called formal power series and, in fact, they arise as qualitative extensions of formal languages. For the domain of weights and their computations, the algebraic structure of semirings proved to be very fruitful (cf. Droste and Kuske [12, page 113]).

The basic concept of weighted automata was first introduced in the seminal paper of Schützenberger [37] in 1961. In particular, he was the first to consider formal power series in the context of Automata Theory. At first, Schützenberger mainly considered the ring of integers. The year after, in [38] he extended his approach to arbitrary rings and, more generally, to semirings. In the beginning, the research on weighted automata and their behaviors was mutually influenced by a close connection to probabilistic automata (cf. [12, page 113]). In 1974, Eilenberg [14] pursued a very general and mathematical approach to Language Theory and Automata Theory and introduced the notion of *multiplicity*:

> "The phenomenon of multiplicity appears whenever any fact takes place for several reasons and we wish to study not only the fact itself but also the reasons for which it takes place. For instance, the fact may be $s \in |\mathcal{A}|$ with $s \in \Sigma^*$ and $\mathcal{A}$ an automaton.[1] The reason for this is the existence of a successful path c in $\mathcal{A}$ with s as label."
>
> Eilenberg [14, page 120]

Classical automata, as introduced in Section 2.2, decide whether or not a given word is recognized, by testing whether there exists a suitable successful path. Hence, classical automata make *qualitative* statements about words. However, as Eilenberg indicates, there may be several successful paths for a specific word. One might be interested in, for instance, *how many* successful paths there are. Thus, we are interested in determining *quantitative* properties of words. This is exactly what weighted automata do. If the underlying semiring is chosen appropriately, then weighted automata are also capable of computing e.g. the resources, time or costs needed for the execution of a specific word.

---

[1] Adapted to our notation, $s \in |\mathcal{A}|$ means $s \in L(\mathcal{A})$, i.e. the automaton $\mathcal{A}$ recognizes the string s.

The main objective in this chapter is to extend the classical notions from Language Theory and Automata Theory, as introduced in Section 2.1 and Section 2.2, to the realm of formal power series. Section 3.1 contains the fundamental theory of semirings and formal power series. We present several examples of semirings that will occur throughout this work. Moreover, we show that formal power series arise as generalizations of languages. The main aim of Section 3.2 is to introduce the notion of weighted automata and thereby extend the concept of recognizability to formal power series. In particular, we show that weighted automata generalize the classical ones from Section 2.2. We conclude this chapter by studying connections between weighted automata and matrices in Section 3.3. These connections give rise to another modeling approach for recognizable power series.

3.1. Semirings and Formal Power Series

While only elementary Graph Theory and Combinatorics were required in the early 1960s, new tools from (non-commutative) Algebra such as semirings and formal power series have been investigated in the context of Automata Theory. Thus, the mathematical foundations of Automata Theory rely on more and more advanced parts of Mathematics, in particular of Algebra (cf. Pin [28, page VII]).

In order to obtain a uniform and powerful model of weighted automata for different realizations of weights and their computations, the underlying weight structures are usually modeled by the algebraic structure of an abstract semiring S. A weighted automaton over Σ with weights coming from a semiring then realizes a map that associates to each word over Σ the total weight of its execution. More precisely, the multiplication of the semiring is used for determining the weights of paths in the automaton, and the weight of a word $w \in \Sigma^*$ is obtained by the sum of all weights of paths labeled by w. In general, a map from Σ^* into a semiring is called a formal power series. Summarizing, the basic foundations for the theory of weighted automata are semirings and formal power series. It is the goal of this section to introduce all the preliminary notions from the theory of semirings and formal power series, as far as they are used in the subsequent sections and chapters. In particular, we give examples of several semirings for different interpretations of weights, which will be used throughout this work. Moreover, we show that formal power series arise as natural generalizations of formal languages.

We set up notation and terminology following Droste and Kuich [9, § 2 and § 3].

Semirings

3.1.1 Definition.

a) A **semiring** is a tuple $(S, +, \cdot, 0, 1)$ where

- $(S, +, 0)$ is a commutative monoid,
- $(S, \cdot, 1)$ is a monoid,

- multiplication distributes over addition from both sides, i.e. the distributivity laws

$$s_1 \cdot (s_2 + s_3) = (s_1 \cdot s_2) + (s_1 \cdot s_3),$$
$$(s_1 + s_2) \cdot s_3 = (s_1 \cdot s_3) + (s_2 \cdot s_3)$$

hold for any $s_1, s_2, s_3 \in S$,

- and 0 annihilates S, i.e.

$$0 \cdot s = s \cdot 0 = 0$$

holds for each $s \in S$.

As for monoids, the tuple $(S, +, \cdot, 0, 1)$ is simply denoted by S if there is no confusion likely to arise.

b) A semiring $(S, +, \cdot, 0, 1)$ is called **commutative** if the multiplicative monoid $(S, \cdot, 1)$ is commutative.

Intuitively, a semiring is a ring (with unity) without the requirement of additive inverses, i.e. without subtraction. Note further that the condition $0 \cdot s = s \cdot 0 = 0$ for any $s \in S$ is not a consequence of the other axioms, as is the case for rings. Since both operations of a semiring are associative, we usually omit parenthesis within sums and products, respectively. Furthermore, we follow the usual convention that multiplication is applied before addition.

3.1.2 Example. A typical example is the (commutative) semiring of natural numbers $(\mathbb{N}, +, \cdot, 0, 1)$, where $+$ and $\cdot$ denote the usual addition and multiplication on $\mathbb{N}$, respectively. Clearly, all rings (with unity) as well as all fields are semirings, e.g. the integers $\mathbb{Z}$, rationals $\mathbb{Q}$, reals $\mathbb{R}$, etc.

3.1.3 Notation. Let S be a semiring.

a) Given $n \in \mathbb{N}$ and $s_1, \ldots, s_n \in S$, we use the following standard abbreviations:

$$\sum_{i=1}^{n} s_i := \begin{cases} 0 & \text{if } n = 0 \\ \sum_{i=1}^{n-1} s_i + s_n & \text{otherwise,} \end{cases}$$

$$\prod_{i=1}^{n} s_i := \begin{cases} 1 & \text{if } n = 0 \\ \prod_{i=1}^{n-1} s_i \cdot s_n & \text{otherwise.} \end{cases}$$

b) Given a finite index set $I = \{i_1, \ldots, i_n\}$ with $n \in \mathbb{N}$ and $s_i \in S$ for $i \in I$, we set:

$$\sum_{i \in I} s_i := \sum_{k=1}^{n} s_{i_k},$$

$$\prod_{i \in I} s_i := \prod_{k=1}^{n} s_{i_k}.$$

As addition in a semiring is commutative, the value of the sum is independent of the enumeration of the index set I. However, the value of the product actually depends on the enumeration of the index set I if S is not commutative. Therefore, when working with products we always need to specify an enumeration of the respective index set.

3.1.4 Definition. Let $(S_1, +_1, \cdot_1, 0_1, 1_1)$ and $(S_2, +_2, \cdot_2, 0_2, 1_2)$ be semirings.

a) A **semiring homomorphism** from S_1 into S_2 is a map $h\colon S_1 \to S_2$ such that

- h is a monoid homomorphism from $(S_1, +_1, 0_1)$ into $(S_2, +_2, 0_2)$ and
- h is a monoid homomorphism from $(S_1, \cdot_1, 1_1)$ into $(S_2, \cdot_2, 1_2)$.

b) A bijective semiring homomorphism is called a **semiring isomorphism.**

c) The semirings S_1 and S_2 are called **isomorphic** if there exists a semiring isomorphism from S_1 into S_2.

As we have already mentioned, different semirings provide different realizations and computations of weights. Next, we consider some specific examples of semirings each of which indicates a different possible understanding of weights.

3.1.5 Example.

a) We have already presented the semiring of non-negative integers $\mathbb{N} = \mathbb{Z}_{\geq 0}$ in Example 3.1.2. Likewise, the non-negative rationals $\mathbb{Q}_{\geq 0} := \{x \in \mathbb{Q} \mid x \geq 0\}$ as well as the non-negative reals $\mathbb{R}_{\geq 0} := \{x \in \mathbb{R} \mid x \geq 0\}$ with their usual operations also form semirings. Later, we will use the semiring $\mathbb{N}$ to count the number of successful paths labeled by a specific word in a given classical automaton.

b) We consider now the extended sets $\overline{\mathbb{N}} := \mathbb{N} \cup \{\infty\}$ and $\overline{\mathbb{R}}_{\geq 0} := \mathbb{R}_{\geq 0} \cup \{\infty\}$. Then one can easily verify that the tuples

$$(\overline{\mathbb{N}}, \min, +, \infty, 0),$$
$$(\overline{\mathbb{R}}_{\geq 0}, \min, +, \infty, 0),$$

where the operations min and $+$ are defined in the obvious fashion, are semirings, called **min-plus semirings** or **tropical semirings**. Analogously, the tuples

$$(\underline{\mathbb{N}}, \max, +, -\infty, 0),$$
$$(\underline{\mathbb{R}}_{\geq 0}, \max, +, -\infty, 0),$$

with $\underline{\mathbb{N}} := \mathbb{N} \cup \{-\infty\}, \underline{\mathbb{R}}_{\geq 0} := \mathbb{R}_{\geq 0} \cup \{-\infty\}$ and max defined in the obvious fashion are semirings, called **max-plus semirings** or **arctic semirings**. Note that addition acts as multiplicative operation both in tropical and in arctic semirings, whereas min and max serve as semiring addition. Furthermore, the semirings $\overline{\mathbb{R}}_{\geq 0}$ and $\underline{\mathbb{R}}_{\geq 0}$ are isomorphic via the semiring isomorphism

$$\overline{\mathbb{R}}_{\geq 0} \to \underline{\mathbb{R}}_{\geq 0}, \; x \mapsto -x.$$

Clearly, the semirings $\overline{\mathbb{N}}$ and $\underline{\mathbb{N}}$ are isomorphic as well. Elements of the tropical and the arctic semirings can be viewed as resources, costs or efforts needed for some action. Using the operations of the semirings we can then compute, for instance, the minimal or maximal costs of a sequence of actions. Therefore, tropic and arctic semirings are often employed in optimization problems, e.g. in network optimization problems (cf. [9, page 7]) and in performance evaluation on discrete event systems.

c) A further example is provided by the **Viterbi semiring** $([0,1], \max, \cdot, 0, 1)$, whose operations are again defined in the obvious fashion and which can be used to compute probabilities.

d) The **Łukasiewicz semiring** $([0,1], \max, \otimes, 0, 1)$ with $x \otimes y = \max(0, x + y - 1)$ for any $x, y \in [0,1]$ occurs in multi-valued logic (cf. [9, page 8] and Droste and Gastin [6, page 70]).

e) The **semiring of (formal) languages** over the alphabet Σ is given by the tuple $(\mathcal{P}(\Sigma^*), \cup, \cdot, \emptyset, \varepsilon)$, where $\cdot$ denotes the language-theoretic concatenation.

Each of these semirings, except the semiring of languages, is commutative.

3.1.6 Example. The **Boolean semiring** $\mathbb{B} = (\{0, 1\}, +, \cdot, 0, 1)$ consists of two distinct elements $0 \neq 1$ and its operations are uniquely determined by the equation

$$1 + 1 = 1.$$

Indeed, the remaining rules of addition $0 + 0 = 0$ and $1 + 0 = 0 + 1 = 1$ are a direct consequence of 0 being the neutral element of addition. Similarly, its multiplication rules $1 \cdot 1 = 1$ and $1 \cdot 0 = 0 \cdot 1 = 0 \cdot 0 = 0$ are entirely forced by the axioms in Definition 3.1.1. The Boolean semiring plays a central role in this work, since it will help us to connect our language-theoretic notions and results from Chapter 2 with the ones that we will introduce and develop in the following.

Formal Power Series

"Weighted automata realise power series—in contrast to 'classical' automata which accept languages. There are many good reasons that make power series worth an interest compared to languages, beyond the raw appeal to generalisation that inhabits every mathematician.
First, power series provide a more powerful mean for modelisation, replacing a pure acceptance/rejection mode by a quantification process. Second, by putting automata theory in a seemingly more complicated framework, one benefits from the strength of mathematical structures thus involved and some results and constructions become simpler, both conceptually, and on the complexity level. Let us also mention, as a third example, that in the beginning of the theory, weighted automata were probably considered for their ability of defining languages—via the supports of realised power series—rather than for the power series themselves."

Sakarovitch [34, page 106]

In general, a formal power series is a map that assigns to each word over Σ an element of a semiring. If the semiring is chosen correspondingly, then its elements may be quantities or weights such as e.g. costs, resources or probabilities. As Sakarovitch indicates, while in the context of classical automata the central objects under consideration are languages, weighted automata realize formal power series by determining quantitative properties of a word (e.g. costs or resources needed for its execution or the number of successful paths). However, it is important to notice that formal power series form a generalization of languages. More precisely, each language can be understood as formal power series by identifying the language with its characteristic series. Conversely, the support of a power series constitutes a language. We infer in particular that there is a bijective correspondence between languages and formal power series with coefficients in the Boolean semiring. For other semirings, formal power series can be viewed as weighted, multi-valued or quantified languages in which each word is assigned a weight, a number or some quantity (cf. Droste and Kuich [9, page 4]). In this sense, formal power series arise as quantitative extension of languages. We conclude this section by drawing a connection between formal power series in the context of Automata Theory and formal power series, as studied in Algebra.

Throughout the rest of this work, let S be a non-trivial semiring with $0 \neq 1$.

3.1.7 Definition. Mappings from Σ^* into S are called **(formal) power series**. The set of all formal power series from Σ^* into S is denoted by $S\langle\langle\Sigma^*\rangle\rangle$.

3.1.8 Notation. For a formal power series $r \in S\langle\langle\Sigma^*\rangle\rangle$ and a word $w \in \Sigma^*$, it is customary to write (r, w) instead of $r(w)$. Moreover, r is often written as a formal sum

$$r = \sum_{w \in \Sigma^*} (r, w)w.$$

The values (r, w) are also referred to as the **coefficients** of the series.

This terminology reflects the intuitive ideas connected with power series. The power series are called *formal* to indicate that we are not interested in summing up the series but rather, for instance, in various operations defined for series (cf. [9, page 12]). We will introduce these operations in Chapter 4. In the literature, the elements of $S\langle\langle\Sigma^*\rangle\rangle$ are often referred to as *formal power series with non-commuting variables* (cf. Salomaa and Soittola [35, § I.3]).

In order to relate formal power series to languages, the following definition is essential.

3.1.9 Definition.

a) Given a formal power series $r \in S\langle\langle\Sigma^*\rangle\rangle$, its **support** is defined to be the language

$$\text{supp}(r) := \{w \in \Sigma^* \mid (r, w) \neq 0\}$$

over the alphabet Σ.

b) Given a language L over Σ, its **characteristic series** is defined to be the formal power series $\mathbb{1}_L \in S\langle\!\langle \Sigma^* \rangle\!\rangle$ given by

$$\mathbb{1}_L(w) := \begin{cases} 1 & \text{if } w \in L \\ 0 & \text{otherwise} \end{cases}$$

for $w \in \Sigma^*$.

c) A formal power series $r \in S\langle\!\langle \Sigma^* \rangle\!\rangle$ is called **unambiguous** if $r(\Sigma^*) \subseteq \{0, 1\}$, i.e. if the series r takes only the values 0 and 1 (cf. Eilenberg [14, page 126]).

Since we assume that $0 \neq 1$ in S, the support of an unambiguous series $r \in S\langle\!\langle \Sigma^* \rangle\!\rangle$ is given by

$$\text{supp}(r) = \{w \in \Sigma^* \mid (r, w) \neq 0\} = \{w \in \Sigma^* \mid (r, w) = 1\}.$$

Thus, the maps

$$\text{char} \colon \mathcal{P}(\Sigma^*) \to S\langle\!\langle \Sigma^* \rangle\!\rangle, \; L \mapsto \mathbb{1}_L,$$

$$\text{supp} \colon S\langle\!\langle \Sigma^* \rangle\!\rangle \to \mathcal{P}(\Sigma^*), \; r \mapsto \text{supp}(r)$$

provide a one-to-one correspondence between the set of languages over Σ and the subset of unambiguous formal power series in $S\langle\!\langle \Sigma^* \rangle\!\rangle$. Therefore, each language can be viewed as an (unambiguous) formal power series with coefficients in S and viceversa each unambiguous formal power series can be understood as a language. Hence, formal power series indeed form a generalization of formal languages, as we have already indicated. If we consider the Boolean semiring $S = \mathbb{B}$, then we even obtain a one-to-one correspondence between the languages over Σ and the set of all formal power series in $\mathbb{B}\langle\!\langle \Sigma^* \rangle\!\rangle$, since every formal power series with coefficients in the Boolean semiring $\mathbb{B}$ is unambiguous. On a set-theoretical level, this correspondence also holds for the ring $\mathbb{F}_2$ of integers modulo 2 instead of the semiring $\mathbb{B}$. In fact, the only difference between the two semirings is the value of the sum $1 + 1$, which is 1 in $\mathbb{B}$ and 0 in $\mathbb{F}_2$. However, we will see e.g. in Proposition 3.2.7 that we do need the arithmetic of $\mathbb{B}$ rather than that of $\mathbb{F}_2$, as it will help us to connect the language-theoretic notions with those concerning formal power series.

For other semirings, formal power series can be viewed as weighted, multi-valued or quantified languages in which each word is assigned a weight, a number, or some quantity (cf. Droste and Kuich [9, page 4]). This indicates that Eilenberg's terminology (cf. [14, § VI.3]), who uses the notion S-*subset of* Σ^* rather than formal power series, is appropriate as well.

3.1.10 Example. Given a classical automaton $\mathcal{A}$ over Σ and a word $w \in \Sigma^*$, we denote by m_w the number of successful paths in $\mathcal{A}$ with label w, which is clearly finite. Then the mapping rule $w \mapsto m_w$ provides a power series with coefficients in the semiring $\mathbb{N}$. Eilenberg appropriately calls the number m_w the *multiplicity with which the word w belongs to* $L(\mathcal{A})$, since each successful path with label w constitutes a "proof" for the fact $w \in L(\mathcal{A})$ (cf. [14, § VI.1]). As a word w is recognized by $\mathcal{A}$ if and only if there exists a successful path carrying w as its label, we obtain that the support of the series

$$m = \sum_{w \in \Sigma^*} m_w w$$

coincides with the language $L(\mathcal{A})$ recognized by the automaton $\mathcal{A}$. Furthermore, we notice that the series $m \in \mathbb{N}\langle\langle \Sigma^* \rangle\rangle$ is unambiguous if and only if the underlying automaton $\mathcal{A}$ is deterministic.

Formal power series are not only a useful concept in the context of Automata Theory, but are also an important notion in Algebra. At first sight, our concept of formal power series does not seem to be completely consistent with that from Algebra. However, by considering a one-letter alphabet we now point out a close connection between the two notions.

3.1.11 Remark. Let S be a semiring and consider the alphabet $\Sigma = \{x\}$. Recall that the monoids $\mathbb{N}$ and $\{x\}^*$ are isomorphic via $n \mapsto x^n$. Therefore, every formal power series $r \in S\langle\langle \{x\}^* \rangle\rangle$ can be written as

$$r = \sum_{w \in \Sigma^*} (r, w)w = \sum_{n \in \mathbb{N}} a_n x^n$$

with $a_n = (r, x^n) \in S$ for $n \in \mathbb{N}$. Thus, a formal power series $r \in S\langle\langle \{x\}^* \rangle\rangle$ can be identified with the sequence $\{(r, x^n)\}_{n \in \mathbb{N}} \subseteq S$ of its coefficients. Consequently, our notion of formal power series over the alphabet $\{x\}$ coincides with the one of ordinary formal power series over S in one variable, as studied in Algebra. Hence, we obtain

$$S\langle\langle \{x\}^* \rangle\rangle = S[[x]],$$

which is a ring if S is a ring (see also Example 4.1.5).

3.2. Weighted Automata and Their Behavior

As we have already indicated at the beginning of this chapter, weighted automata are finite automata whose transitions and states carry weights. Hence, they are capable of determining quantitative properties of words. Formally, a weighted automaton over Σ with weights in S realizes a formal power series over Σ with coefficients in S, called its *behavior*, which assigns to each word the weight of its execution in the automaton. More precisely, we obtain the weight of a word $w \in \Sigma^*$ by taking the sum of all weights of paths labeled by w, and we use the multiplication of S for determining the weights of paths. Thus, the focus lies on the computations that realize a given word, rather than on the qualitative statement whether or not the word is recognized.

The goal of this section is to formally introduce weighted automata and to define how their behavior is determined. In particular, we extend the notion of recognizability to formal power series. We have seen in Section 3.1 that formal power series generalize languages and now we further show that, similarly, weighted automata form a natural generalization of classical automata. Moreover, we present multiple examples to illustrate the expressive power of weighted automata and to point out possible applications. Ultimately, we examine some properties of recognizable power series with respect to certain transformations, which allow us to change the underlying alphabet. We conclude this section by discussing both similarities as well as differences between weighted and classical automata.

The definitions and results in this section are mainly due to Droste and Kuske [12, § 2] and Droste [11, § 5].

3.2.1 Definition. A **weighted finite automaton** over Σ with weights in S is a quadruple $\mathcal{A} = (Q, \text{in}, \text{wt}, \text{out})$ consisting of

- a non-empty finite set Q of **states**,

- a **weight function** in: $Q \to S$ **for entering states**,

- a **transition weight function** wt: $Q \times \Sigma \times Q \to S$, and

- a **weight function** out: $Q \to S$ **for leaving states**.

We often omit the adjective *finite* when we refer to weighted finite automata. Contrary to our terminology with classical automata, we now call every triple (p, a, q) in the Cartesian product $Q \times \Sigma \times Q$ a **transition**. If a transition (p, a, q) is assigned the weight $\text{wt}(p, a, q) = 0$, then it means that the transition is impossible or impracticable. Similarly, it is not possible to enter or leave the automaton in state q if $\text{in}(q) = 0$ or if $\text{out}(q) = 0$, respectively. On the other hand, if we have $\text{in}(q) \neq 0$ or $\text{out}(q) \neq 0$, then the state q can be regarded as initial or final state of the automaton, respectively.

Like classical automata, a weighted automaton $\mathcal{A} = (Q, \text{in}, \text{wt}, \text{out})$ can be illustrated by its **state diagram**:

- Again, the vertices of the state diagram are given by the states of the automaton and are represented by labeled circles.

- Each vertex that corresponds to a state q with $\text{in}(q) \neq 0$ is equipped with an incoming arrow labeled by $\text{in}(q)$, i.e. is depicted as:

$$\text{in}(q) \longrightarrow \boxed{q}$$

If the weight $\text{in}(q)$ for entering the automaton in state q equals 0, then we leave the corresponding vertex unchanged.

- Likewise, each vertex corresponding to a state q with $\text{out}(q) \neq 0$ is equipped with an outgoing arrow labeled by $\text{out}(q)$, i.e. is depicted as:

If the weight $\text{out}(q)$ for leaving the automaton in state q equals 0, then we leave the corresponding vertex unchanged.

- The edges of the state diagram are determined by the transitions that are assigned a non-zero weight. More precisely, each transition (p, a, q) in the automaton with $\mathrm{wt}(p, a, q) \neq 0$ is represented by an edge directed from vertex p to vertex q that is labeled both by the letter a and the transition weight $\mathrm{wt}(p, a, q)$. Pictorially, this means:

$$ p \xrightarrow[\mathrm{wt}(p,a,q)]{a} q $$

If the states p and q coincide, then we equip the vertex p with a loop that is labeled both by a and $\mathrm{wt}(p, a, q)$. Transitions (p, a, q) that are assigned transition weight $\mathrm{wt}(p, a, q) = 0$ do not occur in the state diagram.

As with classical automata, such a state diagram can be used to conversely specify the formal definition of the corresponding automaton.

3.2.2 Example. Consider the tropical semiring $(\overline{\mathbb{N}}, \min, +, \infty, 0)$. The state diagram

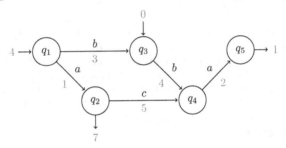

corresponds to the weighted automaton $\mathcal{A} = (Q, \mathrm{in}, \mathrm{wt}, \mathrm{out})$ over Σ and S with state set $Q = \{q_1, q_2, q_3, q_4, q_5\}$ whose weight functions are given by

$$ \mathrm{in}(q) = \begin{cases} 4 & \text{if } q = q_1 \\ 0 & \text{if } q = q_3 \\ \infty & \text{otherwise} \end{cases} \quad (q \in Q), $$

$$ \mathrm{wt}(p, a, q) = \begin{cases} 1 & \text{if } (p, a, q) = (q_1, a, q_2) \\ 3 & \text{if } (p, a, q) = (q_1, b, q_3) \\ 5 & \text{if } (p, a, q) = (q_2, c, q_4) \\ 4 & \text{if } (p, a, q) = (q_3, b, q_4) \\ 2 & \text{if } (p, a, q) = (q_4, a, q_5) \\ \infty & \text{otherwise} \end{cases} \quad (p, q \in Q, a \in \Sigma), $$

$$ \mathrm{out}(q) = \begin{cases} 7 & \text{if } q = q_2 \\ 1 & \text{if } q = q_5 \\ \infty & \text{otherwise} \end{cases} \quad (q \in Q). $$

64

Next, we want to introduce the behavior of a weighted automaton, which can be considered as its semantic representation. Since we modeled our weight structures as semirings, we can give a common definition for the behavior of weighted automata for different realizations of weights and their computations.

3.2.3 Definition. Let $\mathcal{A} = (Q, \text{in}, \text{wt}, \text{out})$ be a weighted automaton over Σ with weights in the semiring S.

a) A **path** in $\mathcal{A}$ is an alternating sequence $P = q_0 a_1 q_1 \ldots a_n q_n \in Q(\Sigma Q)^*$ with $n \in \mathbb{N}$, i.e. P can be regarded as a sequence of transitions. Its **run weight** is the product

$$\text{rweight}(P) := \prod_{i=1}^{n} \text{wt}(q_{i-1}, a_i, q_i).$$

The **weight** of P is then defined by

$$\text{weight}(P) := \text{in}(q_0) \cdot \text{rweight}(P) \cdot \text{out}(q_n).$$

The **label** of P is the word $\text{label}(P) := a_1 \ldots a_n$ over Σ.

b) The **behavior** of $\mathcal{A}$ is the formal power series $||\mathcal{A}|| = \sum_{w \in \Sigma^*} (||\mathcal{A}||, w) w \in S\langle\langle \Sigma^* \rangle\rangle$ whose coefficients are given by

$$(||\mathcal{A}||, w) := \sum_{\substack{P \in \text{Path}(\mathcal{A}) \\ \text{label}(P) = w}} \text{weight}(P)$$

for $w \in \Sigma^*$, where $\text{Path}(\mathcal{A})$ denotes the language $Q(\Sigma Q)^*$ of all paths in $\mathcal{A}$. Thus, the behavior of the weighted automaton $\mathcal{A}$ is a function associating to each word the total weight of its execution in $\mathcal{A}$.

3.2.4 Remark.

a) As we regard every triple (p, a, q) in the Cartesian product $Q \times \Sigma \times Q$ as a transition in the context of weighted automata, the notion of a path in a weighted automaton is slightly different than that in a classical automaton (see Definition 2.2.6). However, the set of paths labeled by a given word $w \in \Sigma^*$ is still finite, as its cardinality is given by

$$|\{P \in \text{Path}(\mathcal{A}) \mid \text{label}(P) = w\}| = |Q|^{|w|+1}$$

and the state set Q is finite. Consequently, $(||\mathcal{A}||, w)$ is a finite sum in S, and hence the behavior of $\mathcal{A}$ is well-defined.

b) The run weight of a path $P = q_0$ consisting of just one state is an empty product and thus given by $\text{rweight}(P) = 1$ (see Notation 3.1.3). Consequently, the weight of such a path is given by the product

$$\text{weight}(P) = \text{in}(q_0) \cdot \text{out}(q_0).$$

In particular, we obtain that the sum

$$(\|\mathcal{A}\|, \varepsilon) = \sum_{q \in Q} \mathrm{in}(q) \, \mathrm{out}(q)$$

is the coefficient corresponding to the empty word.

c) If one of the factors $\mathrm{wt}(q_{i-1}, a_i, q_i)$ ($i = 1, \ldots, n$) equals 0, then the whole product $\mathrm{rweight}(P)$ is annihilated, i.e. we have

$$\mathrm{rweight}(P) = 0.$$

Similarly, if one of the factors $\mathrm{in}(q_0), \mathrm{rweight}(P), \mathrm{out}(q_n)$ equals 0, then the whole product $\mathrm{weight}(P)$ is annihilated, i.e. we obtain

$$\mathrm{weight}(P) = 0.$$

In general, the converse does not hold, since e.g. for the ring $\mathbb{F}_4$ of integers modulo 4 we have $2 \cdot 2 = 0$ even though $2 \neq 0$. However, if in a weighted automaton over an arbitrary semiring all weights of transitions and all weights for entering and leaving states have value 0 or 1, then the weight of a path also evaluates either to 0 or to 1. More precisely, its weight evaluates to 1 if and only if the weights for entering and leaving the path both equal 1, and all its transitions carry weight 1.

d) For each word $w = a_1 \ldots a_n$ over Σ the corresponding coefficient $(\|\mathcal{A}\|, w)$ of the behavior of $\mathcal{A}$ can be rewritten as

$$(\|\mathcal{A}\|, w) = \sum_{q_0, \ldots, q_n \in Q} \mathrm{in}(q_0) \cdot \mathrm{wt}(q_0, a_1, q_1) \cdot \ldots \cdot \mathrm{wt}(q_{n-1}, a_n, q_n) \cdot \mathrm{out}(q_n).$$

3.2.5 Definition. We call a formal power series $r \in S\langle\!\langle \Sigma^* \rangle\!\rangle$ **recognizable** if it is the behavior of some weighted automaton. The set of all recognizable formal power series in $S\langle\!\langle \Sigma^* \rangle\!\rangle$ is denoted by $S^{\mathrm{rec}}\langle\!\langle \Sigma^* \rangle\!\rangle$.

While classical automata provide a qualitative testing procedure, the behaviors of weighted automata comprise computations of weights. However, with these computations we even capture the qualitative method of recognition of classical automata if the underlying weight structure is given by the Boolean semiring $\mathbb{B}$, as explained in the following.

3.2.6 Definition. Given a classical automaton $\mathcal{A} = (Q, I, T, F)$ over Σ, we define a weighted automaton $\mathcal{A}^{\mathrm{w}} = (Q, \mathrm{in}, \mathrm{wt}, \mathrm{out})$ over Σ with the same state set and weights in S by the functions

$$\mathrm{in}(q) = \begin{cases} 1 & \text{if } q \in I \\ 0 & \text{otherwise} \end{cases} \quad (q \in Q),$$

$$\mathrm{wt}(p, a, q) = \begin{cases} 1 & \text{if } (p, a, q) \in T \\ 0 & \text{otherwise} \end{cases} \quad (p, q \in Q, a \in \Sigma),$$

$$\mathrm{out}(q) = \begin{cases} 1 & \text{if } q \in F \\ 0 & \text{otherwise} \end{cases} \quad (q \in Q).$$

3.2.7 Proposition. *The map*

$$\text{char}\colon \mathcal{P}(\Sigma^*) \to \mathbb{B}\langle\!\langle \Sigma^* \rangle\!\rangle, \quad L \mapsto \mathbb{1}_L$$

and its inverse

$$\text{supp}\colon \mathbb{B}\langle\!\langle \Sigma^* \rangle\!\rangle \to \mathcal{P}(\Sigma^*), \quad r \mapsto \text{supp}(r)$$

provide a one-to-one correspondence between the recognizable languages over Σ and the set $\mathbb{B}^{\text{rec}}\langle\!\langle \Sigma^* \rangle\!\rangle$ of recognizable power series from Σ^* into the Boolean semiring $\mathbb{B}$.

Proof. First, we assume that $\mathcal{A} = (Q, I, T, F)$ is a classical automaton over Σ and we consider the weighted automaton $\mathcal{A}^{\text{w}} = (Q, \text{in}, \text{wt}, \text{out})$ over Σ with weights in $\mathbb{B}$ (see Definition 3.2.6). By our considerations in Remark 3.2.4c), we then obtain

$1 = \text{weight}(P) = \text{in}(q_0) \cdot \text{wt}(q_0, a_1, q_1) \cdot \ldots \cdot \text{wt}(q_{n-1}, a_n, q_n) \cdot \text{out}(q_n)$

$\Leftrightarrow$ all factors of the product are non-zero

$\Leftrightarrow$ $q_0 \in I, q_n \in F$ and $(q_{i-1}, a_i, q_i) \in T$ for $i = 1, \ldots, n$

$\Leftrightarrow$ P is a successful path in the classical automaton $\mathcal{A}$ (with label $a_1 \ldots a_n$)

for each path $P = q_0 a_1 q_1 \ldots a_n q_n$ in the weighted automaton $\mathcal{A}^{\text{w}}$. Consequently, exploiting the arithmetic of the Boolean semiring, we obtain

$$1 = (||\mathcal{A}^{\text{w}}||, w) = \sum_{\substack{P \in \text{Path}(\mathcal{A}^{\text{w}}) \\ \text{label}(P) = w}} \text{weight}(P)$$

$\Leftrightarrow$ at least one summand of the sum is non-zero

$\Leftrightarrow$ there exists a $P \in \text{Path}(\mathcal{A}^{\text{w}})$ with $\text{label}(P) = w$ and $\text{weight}(P) = 1$

$\Leftrightarrow$ there exists a successful path in the classical automaton $\mathcal{A}$ with label w

$\Leftrightarrow$ $w \in L(\mathcal{A})$

for each word $w \in \Sigma^*$, yielding

$$||\mathcal{A}^{\text{w}}|| = \mathbb{1}_{L(\mathcal{A})}.$$

This shows that the characteristic series $\mathbb{1}_{L(\mathcal{A})}$ of the language $L(\mathcal{A})$ recognized by the classical automaton $\mathcal{A}$ is a recognizable power series, i.e. $\mathbb{1}_{L(\mathcal{A})} \in \mathbb{B}^{\text{rec}}\langle\!\langle \Sigma^* \rangle\!\rangle$.

To prove conversely that the support of any recognizable power series is a recognizable language, we now assume that $\mathcal{A} = (Q, \text{in}, \text{wt}, \text{out})$ is a weighted automaton over Σ with weights in $\mathbb{B}$, and we define a classical automaton $\mathcal{A}^{\text{c}} = (Q, I, T, F)$ with the same state set by

$$I = \{q \in Q \mid \text{in}(q) = 1\},$$
$$T = \{(p, a, q) \mid \text{wt}(p, a, q) = 1\},$$
$$F = \{q \in Q \mid \text{out}(q) = 1\}.$$

As above, exploiting the arithmetic of the Boolean semiring, we obtain the equivalence

$$w \in L(\mathcal{A}^{c}) \iff (||\mathcal{A}||, w) = 1$$

for each word $w \in \Sigma^*$, yielding

$$L(\mathcal{A}^{c}) = \mathrm{supp}(||\mathcal{A}||).$$

Consequently, the support $\mathrm{supp}(||\mathcal{A}||)$ of the series $||\mathcal{A}|| \in \mathbb{B}^{\mathrm{rec}}\langle\langle\Sigma^*\rangle\rangle$ is a recognizable language, completing the proof. □

Thus, the structure of the Boolean semiring $\mathbb{B}$ perfectly models the binary behavior of classical automata. More precisely, classical automata correspond precisely to weighted automata with weights in $\mathbb{B}$. In general, assuming that S is an arbitrary semiring, weighted automata generalize the classical ones in the following sense:

3.2.8 Proposition. *The injective map*

$$\mathrm{char} \colon \mathcal{P}(\Sigma^*) \to S\langle\langle\Sigma^*\rangle\rangle, \ L \mapsto \mathbb{1}_L$$

preserves recognizability for any semiring S, i.e. if $L \subseteq \Sigma^*$ is a recognizable language, then its characteristic series $\mathbb{1}_L$ belongs to $S^{\mathrm{rec}}\langle\langle\Sigma^*\rangle\rangle$ for any semiring S.

Proof. Let L be a recognizable language over Σ. Then there exists a deterministic automaton $\mathcal{A} = (Q, I, T, F)$ over Σ such that $L(\mathcal{A}) = L$ (see Remark 2.2.8). We now consider the weighted automaton $\mathcal{A}^{\mathrm{w}} = (Q, \mathrm{in}, \mathrm{wt}, \mathrm{out})$ over Σ with weights in S (see Definition 3.2.6). Then for every path the weights for entering and leaving the path and the weights of its transitions evaluate either to 0 or to 1. Thus, also the product $\mathrm{weight}(P)$ either takes value 0 or value 1 in S (see Remark 3.2.4c)). Moreover, since we assume the classical automaton $\mathcal{A}$ to be deterministic, for each word $w \in \Sigma^*$ there is at most one successful path in $\mathcal{A}$ with label w. Consequently, there is at most one path in $\mathcal{A}^{\mathrm{w}}$ with label w and whose weight is non-zero. More precisely, such a (successful) path exists if and only if $w \in L(\mathcal{A})$. Therefore, for each word $w \in \Sigma^*$ we obtain

$$(||\mathcal{A}^{\mathrm{w}}||, w) = \sum_{\substack{P \in \mathrm{Path}(\mathcal{A}^{\mathrm{w}}) \\ \mathrm{label}(P) = w}} \underbrace{\mathrm{weight}(P)}_{\in \{0,1\}}$$

$$= \begin{cases} 1 & \text{if } w \in L(\mathcal{A}) \\ 0 & \text{otherwise.} \end{cases}$$

Hence, the behavior $||\mathcal{A}^{\mathrm{w}}||$ of the weighted automaton $\mathcal{A}^{\mathrm{w}}$ is unambiguous and fulfills both $||\mathcal{A}^{\mathrm{w}}|| = \mathbb{1}_{L(\mathcal{A})}$ and $\mathrm{supp}(||\mathcal{A}^{\mathrm{w}}||) = L(\mathcal{A})$. In particular, we achieve

$$\mathbb{1}_L = \mathbb{1}_{L(\mathcal{A})} = ||\mathcal{A}^{\mathrm{w}}|| \in S^{\mathrm{rec}}\langle\langle\Sigma^*\rangle\rangle,$$

completing the proof. □

Generalizing one direction of Proposition 3.2.7, Proposition 3.2.8 implies that every recognizable language can be regarded as a recognizable (unambiguous) power series in $S^{\mathrm{rec}}\langle\!\langle\Sigma^*\rangle\!\rangle$ for any semiring S (by identifying it with its characteristic series). Hence, for unambiguous power series the notion of recognizability is independent of the underlying semiring.

3.2.9 Remark. In the proof of Proposition 3.2.8, it is crucial that we suppose the classical automaton $\mathcal{A}$ to be deterministic. Indeed, if $\mathcal{A}$ is not deterministic, then the behavior of the associated weighted automaton $\mathcal{A}^{\mathrm{w}}$ is not unambiguous and particularly does in general not fulfill $||\mathcal{A}^{\mathrm{w}}|| = \mathbb{1}_{L(\mathcal{A})}$. For instance, we can consider the classical automaton $\mathcal{A}$ given by the state diagram:

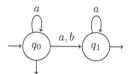

There are exactly three paths in $\mathcal{A}$ that realize the letter a. Thus, the associated weighted automaton $\mathcal{A}^{\mathrm{w}}$ over the semiring $\mathbb{N}$ of natural numbers fulfills

$$(||\mathcal{A}^{\mathrm{w}}||, a) = \sum_{\substack{P \in \mathrm{Path}(\mathcal{A}^{\mathrm{w}}) \\ \mathrm{label}(P)=a}} \mathrm{weight}(P)$$

$$= \underbrace{\mathrm{weight}(q_0 a q_0)}_{=1} + \underbrace{\mathrm{weight}(q_0 a q_1)}_{=1} + \underbrace{\mathrm{weight}(q_1 a q_1)}_{=0}$$

$$= 2 \notin \{0, 1\}.$$

However, one can prove that the support of every series in $\mathbb{N}^{\mathrm{rec}}\langle\!\langle\Sigma^*\rangle\!\rangle$ is a recognizable language (cf. Berstel und Reutenauer [2, Chapter III, Lemma 1.3]). Consequently, a language is recognizable if and only if it is the support of some recognizable series with coefficients in $\mathbb{N}$. Further results regarding the connection between languages and power series with respect to recognizability can be found in [2, Chapter III] and in Salomaa and Soittola [35, §II.5]. Actually, both deal with rationality rather than recognizability, but in Chapter 4 we will exhibit that the two notions coincide.

To illustrate the diversity of weighted automata, we consider again the semirings from Example 3.1.5, which provide different interpretations of weights and thus different application scenarios for weighted automata and their behaviors (cf. Droste, Kuich and Vogler [8, page v] and Droste [11, pages 55–59]).

3.2.10 Example.

a) While classical automata determine whether a given word is recognized or not, weighted automata over the semiring $(\mathbb{N}, +, \cdot, 0, 1)$ of natural numbers allow us to count the number of successful paths labeled by a specific word (cf. Droste and Kuske [12, page 116]).

More precisely, let $\mathcal{A} = (Q, I, T, F)$ be a classical automaton and consider the power series $m \in \mathbb{N}\langle\langle\Sigma^*\rangle\rangle$ associating to each word $w \in \Sigma^*$ the multiplicity m_w with which it belongs to $L(\mathcal{A})$, i.e. the number of successful paths in $\mathcal{A}$ having w as their label. We already encountered this formal power series in Example 3.1.10, where we argued that the support of this series coincides with the recognizable language $L(\mathcal{A})$. Our aim is now to show that already the series m itself is recognizable, i.e. that we have $m \in \mathbb{N}^{rec}\langle\langle\Sigma^*\rangle\rangle$. To this end, we consider the weighted automaton $\mathcal{A}^w = (Q, \text{in}, \text{wt}, \text{out})$ with weights in $\mathbb{N}$ (see Definition 3.2.6). As in the proof of Proposition 3.2.8, we obtain

$$\text{weight}(P) = \begin{cases} 1 & \text{if } P \text{ is a successful path in } \mathcal{A} \\ 0 & \text{otherwise} \end{cases}$$

for each path P in $\mathcal{A}^w$ and consequently

$$(||\mathcal{A}^w||, w) = \sum_{\substack{P \in \text{Path}(\mathcal{A}^w) \\ \text{label}(P)=w}} \text{weight}(P) = m_w$$

for each word $w \in \Sigma^*$. Thus, the behavior of the weighted automaton $\mathcal{A}^w$ associated to $\mathcal{A}$ coincides with the formal power series m, which implies $m \in \mathbb{N}^{rec}\langle\langle\Sigma^*\rangle\rangle$.

b) Next, we consider a weighted automaton $\mathcal{A} = (Q, \text{in}, \text{wt}, \text{out})$ with weights taken from the tropical semiring $\text{Trop} = (\overline{\mathbb{R}}_{\geq 0}, \min, +, \infty, 0)$. We recall that min, with neutral element ∞, acts as addition and $+$, having 0 as neutral element, acts as multiplication of the semiring Trop. Thus, we consider a transition (p, a, q) to be impracticable or impossible if $\text{wt}(p, a, q) = \infty$. Furthermore, the weight of a path $P = q_0 a_1 q_1 \ldots a_n q_n$ in $\mathcal{A}$ is given by the sum

$$\text{weight}(P) = \text{in}(q_0) + \underbrace{\text{wt}(q_0, a_1, q_1) + \cdots + \text{wt}(q_{n-1}, a_n, q_n)}_{= \text{rweight}(P)} + \text{out}(q_n),$$

i.e. all weights along the path P are accumulated. The coefficient of $||\mathcal{A}||$ corresponding to $w \in \Sigma^*$ is determined by the expression

$$(||\mathcal{A}||, w) = \min_{\substack{P \in \text{Path}(\mathcal{A}) \\ \text{label}(P)=w}} \text{weight}(P),$$

i.e. for each word the best or less costly path is chosen. Now, we can understand words as processes (i.e. the letters represent certain actions) and paths in $\mathcal{A}$ as executions or realizations of their labels (cf. Example 2.2.5). Moreover, the weights in $\mathcal{A}$ can be viewed as costs, e.g. each transition can carry as weight the amount of resources or time needed for its execution. Then the behavior $||\mathcal{A}||$ determines for each process the minimal costs needed for its execution in the quantitative system given by $\mathcal{A}$. We could also replace the costs by profits and then be interested in the maximal profit realized by a given process. We can model this scenario by changing the semiring and consider the arctical semiring $\text{Arc} = (\overline{\mathbb{R}}_{\geq 0}, \max, +, -\infty, 0)$ instead.

Weighted automata can not only be used for determining minimal or maximal, but also for determining average costs (cf. Droste [11, page 60]).

c) If we consider the Viterbi semiring Prob $= ([0, 1], \max, \cdot, 0, 1)$, then a weighted automaton $\mathcal{A} = (Q, \mathrm{in}, \mathrm{wt}, \mathrm{out})$ over Prob can be regarded as a classical automaton whose initial states, final states and transitions carry probabilities. For instance, the weight of a transition can be viewed as its reliability, i.e. the probability that the transition occurs correctly. As usual, the weight of a path $P = q_0 a_1 q_1 \ldots a_n q_n$ in $\mathcal{A}$ is given by the product

$$\mathrm{weight}(P) = \mathrm{in}(q_0) \cdot \underbrace{\mathrm{wt}(q_0, a_1, q_1) \cdot \ldots \cdot \mathrm{wt}(q_{n-1}, a_n, q_n)}_{= \, \mathrm{rweight}(P)} \cdot \mathrm{out}(q_n),$$

which then can be understood as the reliability of the path. Moreover, the coefficient

$$(||\mathcal{A}||, w) = \max_{\substack{P \in \mathrm{Path}(\mathcal{A}) \\ \mathrm{label}(P) = w}} \mathrm{weight}(P)$$

of the behavior corresponding to a word $w \in \Sigma^*$ expresses the maximal reliability of its realization by a successful path. This close relationship of weighted automata to probability, in particular to probabilistic automata, led to various applications (cf. e.g. Baier, Größer and Ciesinski [1]).

d) Rahonis [32] deals with the relationship of weighted automata to multi-valued logic, more precisely to fuzzy logic and fuzzy languages. In particular, he considers Łukasiewicz logic and weighted automata over the Łukasiewicz semiring $([0, 1], \max, \otimes, 0, 1)$ with $x \otimes y = \max(0, x + y - 1)$ for $x, y \in [0, 1]$. In general, fuzzy logic is a multi-valued logic over the interval $[0, 1] \subseteq \mathbb{R}$ enriched with further quantifiers like *most, few, many* and *several*. Usually, the underlying structures for the computation of probabilities are modeled by so-called *lattices*, which under certain assumptions constitute semirings. The fuzzy automaton model is a natural *fuzzification* of the classical finite automaton and it is actually a weighted automaton model. Fuzzy structures and fuzzy logic contribute to a wide range of real world applications, since they can effectively incorporate the impreciseness of practical problems (cf. [32, page 482]).

e) Considering weighted automata over Σ with weights coming from the semiring $(\mathcal{P}(\Gamma^*), \cup, \cdot, \emptyset, \{\varepsilon\})$ of languages over the alphabet Γ, we even capture the important notion of a transducer (cf. Droste and Kuske [12, page 116]). The behavior of such an automaton is a map that associates to each word over Σ a language over Γ.

Summarizing, weighted automata determine quantitative properties while classical automata provide only qualitative statements. Moreover, the choice of the underlying semiring admits many different interpretations and applications of (the behavior of) weighted automata. For further examples and applications we refer to Droste, Kuich and Vogler [8], Droste and Kuske [12, page 114 f.] and Droste [11, pages 57–60].

Until now, we have dealt with arbitrary weighted automata over certain semirings. In the following examples, inspired by Eilenberg [14, page 121 f.], we consider specific weighted automata over the semiring $\mathbb{N}$ of natural numbers.

3.2.11 Example. Consider the alphabet $\Sigma = \{a, b\}$ and the classical automaton $\mathcal{A}$ given by the state diagram:

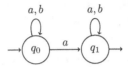

By our considerations in Example 3.2.10a), the weighted automaton $\mathcal{A}^w$ over the semiring $(\mathbb{N}, +, \cdot, 0, 1)$ fulfills $\mathrm{supp}(\|\mathcal{A}^w\|) = L(\mathcal{A}) = \Sigma^*\{a\}\Sigma^*$. More precisely, its behavior is given by the formal power series

$$\|\mathcal{A}^w\| = \sum_{w \in \Sigma^*} m_w w$$

where m_w is the number of successful paths in $\mathcal{A}$ with label w. Now for this particular automaton we have

$$m_w = |w|_a$$

for each $w \in \Sigma^*$, where $|w|_a$ denotes the number of occurrences of the letter a in the word w. Indeed, each occurrence of the letter a corresponds to a possibility to take the transition (q_0, a, q_1) in $\mathcal{A}$, i.e. to pass from the only initial state to the only final state. Consequently, we obtain

$$\|\mathcal{A}^w\| = \sum_{w \in \Sigma^*} m_w w = \sum_{w \in \Sigma^*} |w|_a w,$$

and hence the series $|\cdot|_a \colon \Sigma^* \to \mathbb{N}$, $w \mapsto |w|_a$ is recognizable.

3.2.12 Example. Consider the alphabet $\{x\}$ and the classical automaton $\mathcal{A}$ given by the state diagram:

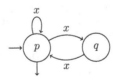

Again by our considerations in Example 3.2.10a), the weighted automaton $\mathcal{A}^w$ over the semiring $(\mathbb{N}, +, \cdot, 0, 1)$ has as its behavior the formal power series

$$\|\mathcal{A}^w\| = \sum_{n \in \mathbb{N}} a_n x^n$$

where the coefficient a_n is given by the number of successful paths in $\mathcal{A}$ with label x^n.

Clearly, we have

$$a_0 = a_1 = 1.$$

We now consider an arbitrary $n \in \mathbb{N}$ with $n \geq 2$. A successful path in $\mathcal{A}$ with label x^n has to end in the only final state p, and thus either realizes the last letter by the loop (p, x, p) or by the transition (q, x, p), where the latter implies in particular that the penultimate letter is realized by the transition (p, x, q). Thus, such a successful path has either the form

$$P_1 x p$$

or the form

$$P_2 x q x p$$

where P_1 and P_2 are successful paths with labels x^{n-1} and x^{n-2}, respectively. This implies

$$a_n = a_{n-1} + a_{n-2}$$

for each $n \geq 2$. Hence, the coefficients of the behavior $||\mathcal{A}^w||$ form the Fibonacci sequence.

For the rest of this section, we follow Droste [11, § 7] to define transformations on formal power series induced by monoid homomorphisms of the form $h \colon \Sigma^* \to \Gamma^*$, where Σ and Γ are alphabets.

3.2.13 Definition. Let Σ, Γ be alphabets and $h \colon \Sigma^* \to \Gamma^*$ a monoid homomorphism.

a) We define a map $h^{-1} \colon S\langle\!\langle \Gamma^* \rangle\!\rangle \to S\langle\!\langle \Sigma^* \rangle\!\rangle$ by setting $h^{-1}(r) := r \circ h$ for $r \in S\langle\!\langle \Gamma^* \rangle\!\rangle$, i.e. we have

$$(h^{-1}(r), w) = (r, h(w))$$

for $w \in \Sigma^*$.

b) If h is non-deleting, then we define a map $h \colon S\langle\!\langle \Sigma^* \rangle\!\rangle \to S\langle\!\langle \Gamma^* \rangle\!\rangle$ by setting

$$(h(r), v) := \sum_{w \in h^{-1}(v)} (r, w)$$

for $r \in S\langle\!\langle \Sigma^* \rangle\!\rangle$ and $v \in \Gamma^*$. As the homomorphism h is assumed to be non-deleting, the preimage $h^{-1}(v)$ is a finite set for any $v \in \Gamma^*$, and hence the map h is well-defined.

3.2.14 Remark. Let Σ, Γ be alphabets and $h \colon \Sigma^* \to \Gamma^*$ a monoid homomorphism.

a) If we work with the symbol h^{-1}, then it is usually clear from the context whether we mean the preimage of h or the map regarding series. In fact, considering the Boolean semiring $S = \mathbb{B}$ we even obtain the correspondence

$$h^{-1}(\mathbb{1}_L) = \mathbb{1}_{h^{-1}(L)}$$

for every language L over Γ.

Indeed, for each $w \in \Sigma^*$ we have

$$1 = (h^{-1}(\mathbb{1}_L), w) = (\mathbb{1}_L, h(w)) \iff h(w) \in L$$
$$\iff w \in h^{-1}(L)$$
$$\iff (\mathbb{1}_{h^{-1}(L)}, w) = 1.$$

b) Likewise, if we use the symbol h, then it is usually clear from the context whether we mean the language-theoretic map or the map regarding series. Furthermore, considering the Boolean semiring $S = \mathbb{B}$, we obtain the correspondence

$$h(\mathbb{1}_L) = \mathbb{1}_{h(L)}$$

for every language L over Σ. Indeed, exploiting the arithmetic of the Boolean semiring, for each $v \in \Gamma^*$ we have

$$1 = (h(\mathbb{1}_L), v) = \sum_{w \in h^{-1}(v)} (\mathbb{1}_L, w) \iff w \in L \text{ for some } w \in h^{-1}(v)$$
$$\iff v \in h(L)$$
$$\iff (\mathbb{1}_{h(L)}, v) = 1,$$

since a sum in the Boolean semiring has value 1 if and only if one of the summands equals 1.

c) In Chapter 4 we define operations on formal power series making the set $S\langle\!\langle \Sigma^* \rangle\!\rangle$ a semiring. Then one can show that the transformations h^{-1} and h constitute semiring homomorphisms with respect to these operations (see Proposition 4.1.8 and Proposition 4.1.9).

Our observations in Remark 3.2.14 show that the transformations of power series from Definition 3.2.13 correspond to taking the preimage and image of languages under monoid homomorphisms from Σ^* into Γ^*. The following results, which are proved in [11, Theorem 7.1 and Theorem 7.3], show that these changes of the underlying alphabet preserve the recognizability of power series. Thus, they constitute generalizations of Proposition 2.2.10 and Proposition 2.2.11.

3.2.15 Proposition. *Let Σ, Γ be alphabets and $h \colon \Sigma^* \to \Gamma^*$ a monoid homomorphism. The map*

$$h^{-1} \colon S\langle\!\langle \Gamma^* \rangle\!\rangle \to S\langle\!\langle \Sigma^* \rangle\!\rangle$$

preserves recognizability, i.e. if $r \in S^{\mathrm{rec}}\langle\!\langle \Gamma^* \rangle\!\rangle$, then the series

$$h^{-1}(r) = r \circ h = \sum_{w \in \Sigma^*} (r, h(w)) w$$

belongs to $S^{\mathrm{rec}}\langle\!\langle \Sigma^* \rangle\!\rangle$.

3.2.16 Proposition. *Let* Σ, Γ *be alphabets and* $h \colon \Sigma^* \to \Gamma^*$ *a non-deleting monoid homomorphism. The map*

$$h \colon S\langle\!\langle \Sigma^* \rangle\!\rangle \to S\langle\!\langle \Gamma^* \rangle\!\rangle$$

preserves recognizability, i.e. if $r \in S^{\mathrm{rec}}\langle\!\langle \Sigma^* \rangle\!\rangle$, *then the series*

$$h(r) = \sum_{v \in \Gamma^*} \left(\sum_{w \in h^{-1}(v)} (r, w) \right) v$$

belongs to $S^{\mathrm{rec}}\langle\!\langle \Gamma^* \rangle\!\rangle$.

In [11, Theorem 7.3], Droste actually considers rationality rather than recognizability, but we will see in Chapter 4 that the two notions coincide.

3.2.17 Remark. Let Σ, Γ be alphabets and $h \colon \Sigma^* \to \Gamma^*$ a monoid homomorphism.

a) Let L be a language over Γ. Then we have $h^{-1}(\mathbb{1}_L) = \mathbb{1}_{h^{-1}(L)}$ in $\mathbb{B}\langle\!\langle \Sigma^* \rangle\!\rangle$ (see Remark 3.2.14a)). Thus, by Proposition 3.2.7 and Proposition 3.2.15 we obtain

$$L \text{ is recognizable } \Leftrightarrow \mathbb{1}_L \in \mathbb{B}^{\mathrm{rec}}\langle\!\langle \Gamma^* \rangle\!\rangle$$
$$\Rightarrow h^{-1}(\mathbb{1}_L) = \mathbb{1}_{h^{-1}(L)} \in \mathbb{B}^{\mathrm{rec}}\langle\!\langle \Sigma^* \rangle\!\rangle$$
$$\Leftrightarrow h^{-1}(L) \text{ is recognizable.}$$

Hence, Proposition 3.2.15 indeed generalizes Proposition 2.2.10.

b) Let L be a language over Σ. If we assume that h is non-deleting, then we have $h(\mathbb{1}_L) = \mathbb{1}_{h(L)}$ in $\mathbb{B}\langle\!\langle \Gamma^* \rangle\!\rangle$ (see Remark 3.2.14b)). Thus, by Proposition 3.2.7 and Proposition 3.2.15 we obtain

$$L \text{ is recognizable } \Leftrightarrow \mathbb{1}_L \in \mathbb{B}^{\mathrm{rec}}\langle\!\langle \Sigma^* \rangle\!\rangle$$
$$\Rightarrow h(\mathbb{1}_L) = \mathbb{1}_{h(L)} \in \mathbb{B}^{\mathrm{rec}}\langle\!\langle \Gamma^* \rangle\!\rangle$$
$$\Leftrightarrow h(L) \text{ is recognizable.}$$

Hence, Proposition 3.2.16 indeed generalizes Proposition 2.2.11.

Until now, we have encountered various similarities between classical and weighted automata. Indeed, many constructions, algorithms and results known from classical Automata Theory can be transferred and performed very generally for weighted automata over large classes of semirings. Thus, many properties of languages regarding recognizability transfer also to formal power series (in some cases even with very similar proofs). For particular properties, sometimes additional assumptions on the underlying semiring are needed (cf. Droste, Kuich and Vogler [8, page vi]). However, there are also differences between classical and finite automata (cf. Droste and Kuske [12, page 116 f.]). For instance, not every recognizable power series can be expressed as the behavior of some *deterministic* weighted automaton (cf. [12, Example 2.5]). Furthermore, the decidability properties of classical automata from Theorem 2.2.14 do in general not transfer to weighted automata. For instance, the Equivalence Problem in the context of weighted automata, i.e. the question whether the behaviors of two given weighted automata do coincide, is undecidable for tropical and arctic semirings (cf. [12, page 136 f.]).

3.3. Linear Representations

In the context of classical automata, the transition set can be represented by the *adjacency matrix* of the corresponding state diagram, which then can be used to compute the language recognized by the automaton (cf Droste, Kuich and Vogler [8, Chapter 3, Example 2.1]). This indicates a close connection between automata and matrices. Indeed, weighted automata can also be represented in terms of matrices. More precisely, the behavior of a weighted automaton can be compactly described using matrices whose entries are elements of the underlying semiring, namely the transition weights of the weighted automaton. Hence, results and methods from Linear Algebra can be used to derive properties of weighted automata. In particular, the more algebraic formalism of matrices often yields very concise, elegant and convincing proofs about recognizable power series. This algebraic approach will be very helpful in proving the Kleene–Schützenberger Theorem (see Section 4.3).

Based on Kuich and Salomaa [21, § I.4], we start by introducing the basic foundations in the context of matrices, as far as they are used in the following.

3.3.1 Definition. Let Q_1 and Q_2 be non-empty finite sets.

a) A map $M \colon Q_1 \times Q_2 \to S$ is called a **matrix**. Given $p \in Q_1$ and $q \in Q_2$, we refer to the value $M(p,q)$ as (p,q)–**entry** or just as **entry** of M, and we denote it by $M_{p,q}$.

b) The set $S^{Q_1 \times Q_2}$ of maps from $Q_1 \times Q_2$ into S comprises all matrices with entries in the semiring S that are indexed by the sets Q_1 and Q_2.

c) If Q_1 or Q_2 is a singleton, then we denote the set $S^{Q_1 \times Q_2}$ simply by S^{Q_1} or S^{Q_2} and refer to its elements as **row vectors** or **column vectors**, respectively. Given a set $Q \in \{Q_1, Q_2\}$, we denote the entries of a row or column vector $V \in S^Q$, respectively, by V_q for any $q \in Q$.

d) We usually identify a matrix $M \in S^{Q_1 \times Q_2}$ with its entry if both Q_1 and Q_2 are singletons.

e) For matrices $M, N \in S^{Q_1 \times Q_2}$, we define their **sum** $M + N \in S^{Q_1 \times Q_2}$ as usual by

$$(M + N)_{p,q} := M_{p,q} + N_{p,q}$$

for any $p \in Q_1$ and $q \in Q_2$.

f) Let Q_3 be a further non-empty finite set. We define the **product** $M \cdot N \in S^{Q_1 \times Q_3}$ of matrices $M \in S^{Q_1 \times Q_2}$ and $N \in S^{Q_2 \times Q_3}$ as usual by

$$(M \cdot N)_{p,r} := \sum_{q \in Q_2} M_{p,q} \cdot N_{q,r}$$

for any $p \in Q_1$ and $r \in Q_3$.

g) Let Q be a non-empty finite set. The **unit matrix** $E_Q \in S^{Q \times Q}$ is defined by

$$(E_Q)_{p,q} := \begin{cases} 1 & \text{if } p = q \\ 0 & \text{otherwise} \end{cases}$$

for any $p, q \in Q$.

3.3.2 Remark. Let Q be a non-empty finite set. Since in semirings, multiplication distributes over addition from both sides, the above defined matrix multiplication is associative. Thus, as the unit matrix E_Q acts as neutral element of matrix multiplication in $S^{Q \times Q}$, the tuple $(S^{Q \times Q}, \cdot, E_Q)$ constitutes a monoid. With the above defined pointwise addition of matrices and the zero matrix $N_Q \in S^{Q \times Q}$, whose entries all have value 0, the tuple $(S^{Q \times Q}, +, \cdot, N_Q, E_Q)$ even forms a semiring. If $|Q| \geq 2$, then the semiring $S^{Q \times Q}$ of matrices is not commutative, even if the underlying semiring S is commutative (cf. [21, page 41]).

3.3.3 Notation. Since the index sets Q_1 and Q_2 of matrices are finite, we can enumerate them as $Q_1 = \{p_0, \ldots, p_n\}$ and $Q_2 = \{q_0, \ldots, q_m\}$ with $n, m \in \mathbb{N}$. Then we can represent a matrix $M \in S^{Q_1 \times Q_2}$ by the following rectangular array:

$$\begin{pmatrix} M_{p_0,q_0} & M_{p_0,q_1} & \cdots & M_{p_0,q_m} \\ M_{p_1,q_0} & M_{p_1,q_1} & \cdots & M_{p_1,q_m} \\ \vdots & \vdots & \ddots & \vdots \\ M_{p_n,q_0} & M_{p_n,q_1} & \cdots & M_{p_n,q_m} \end{pmatrix}$$

In the following, we aim at representing weighted automata in terms of matrices. To this end, we follow Droste and Kuske [12, § 3] and Droste [11, pages 65–68].

3.3.4 Notation. We write $P \colon p \xrightarrow{w}_{\mathcal{A}} q$ for "P is a path in the weighted automaton $\mathcal{A}$ with label w starting in state p and leading to state q" and we omit the index $\mathcal{A}$ if no confusion is likely to arise.

3.3.5 Lemma. *Let $\mathcal{A} = (Q, \text{in}, \text{wt}, \text{out})$ be a weighted automaton over Σ with weights in S. The map*

$$\mu \colon \Sigma^* \to S^{Q \times Q},$$

defined by

$$\mu(w)_{p,q} = \sum_{P \colon p \xrightarrow{w}_{\mathcal{A}} q} \text{rweight}(P) \tag{1}$$

for $p, q \in Q$ and $w \in \Sigma^*$, constitutes a monoid homomorphism from Σ^* into the multiplicative monoid $(S^{Q \times Q}, \cdot, E_Q)$ of matrices.

Proof. We first show that we have $\mu(\varepsilon) = E_Q$. Given distinct states $p \neq q$ in Q, there is no path with label ε starting in p and leading to q. On the other hand, for each $q \in Q$, the unique path P with label ε that leads from state q to q itself is given by $P = q$.

Consequently, we obtain

$$\mu(\varepsilon)_{p,q} = \sum_{P:\, p \xrightarrow{\varepsilon} q} \mathrm{rweight}(P) = \begin{cases} \mathrm{rweight}(q) = 1 & \text{if } p = q \\ \sum_{P \in \emptyset} \mathrm{rweight}(P) = 0 & \text{otherwise} \end{cases}$$

for $p, q \in Q$ (see Remark 3.2.4b)), yielding $\mu(\varepsilon) = E_Q$.

We now show that $\mu(wv) = \mu(w) \cdot \mu(v)$ for any $w, v \in \Sigma^*$. Every path with label wv can be considered as a concatenation of paths P_w and P_v with labels w and v, respectively. More precisely, if $P = q_0 a_1 q_1 \ldots a_n q_n$ is a path with label wv and w has length $|w| = k$, then $P' = q_0 a_1 q_1 \ldots a_k q_k$ is a path with label w and $P'' = q_k a_{k+1} q_{k+1} \ldots a_n q_n$ is a path with label v, and we have $\mathrm{rweight}(P) = \mathrm{rweight}(P') \cdot \mathrm{rweight}(P'')$. Thus, exploiting the distributivity in the semiring S, we achieve:

$$\mu(wv)_{p,q} = \sum_{P:\, p \xrightarrow{wv} q} \mathrm{rweight}(P)$$

$$= \sum_{r \in Q} \sum_{P':\, p \xrightarrow{w} r} \sum_{P'':\, r \xrightarrow{v} q} \mathrm{rweight}(P') \cdot \mathrm{rweight}(P'')$$

$$= \sum_{r \in Q} \left(\left(\sum_{P':\, p \xrightarrow{w} r} \mathrm{rweight}(P') \right) \cdot \left(\sum_{P'':\, r \xrightarrow{v} q} \mathrm{rweight}(P'') \right) \right)$$

$$= \sum_{r \in Q} \mu(w)_{p,r} \cdot \mu(v)_{r,q}$$

$$= (\mu(w) \cdot \mu(v))_{p,q}$$

$\square$

Hence, the weight function wt of a weighted automaton $\mathcal{A} = (Q, \mathrm{in}, \mathrm{wt}, \mathrm{out})$ indicates a monoid homomorphisms μ that associates to each word $w \in \Sigma^*$ a matrix $\mu(w) \in S^{Q \times Q}$. If we further define a row vector $\lambda \in S^Q$ by

$$\lambda_q = \mathrm{in}(q) \tag{2}$$

and a column vector $\gamma \in S^Q$ by

$$\gamma_q = \mathrm{out}(q) \tag{3}$$

for $q \in Q$, then we can rewrite the behavior of $\mathcal{A}$ in terms of matrices as follows:

3.3.6 Lemma. *Let $\mathcal{A} = (Q, \mathrm{in}, \mathrm{wt}, \mathrm{out})$ be a weighted automaton over Σ with weights in S. Then the coefficients of its behavior are given by*

$$(\|\mathcal{A}\|, w) = \sum_{p,q \in Q} \lambda_p \cdot \mu(w)_{p,q} \cdot \gamma_q = \lambda \cdot \mu(w) \cdot \gamma \tag{4}$$

for $w \in \Sigma^*$, where $\mu \colon \Sigma^* \to S^{Q \times Q}$ denotes the monoid homomorphism defined by (1), and the vectors $\lambda, \gamma \in S^Q$ are defined by (2) and (3).

Proof. Exploiting the distributivity in S, we obtain

$$
\begin{aligned}
(||\mathcal{A}||, w) &= \sum_{\substack{P \in \mathrm{Path}(\mathcal{A}) \\ \mathrm{label}(P) = w}} \mathrm{weight}(P) = \sum_{p,q \in Q} \sum_{P\,:\, p \xrightarrow{w} q} \mathrm{weight}(P) \\
&= \sum_{p,q \in Q} \sum_{P\,:\, p \xrightarrow{w} q} \mathrm{in}(p) \cdot \mathrm{rweight}(P) \cdot \mathrm{out}(q) \\
&= \sum_{p,q \in Q} \mathrm{in}(p) \cdot \left(\sum_{P\,:\, p \xrightarrow{w} q} \mathrm{rweight}(P) \right) \cdot \mathrm{out}(q) \\
&= \sum_{p,q \in Q} \lambda_p \cdot \mu(w)_{p,q} \cdot \gamma_q \\
&= \lambda \cdot \mu(w) \cdot \gamma
\end{aligned}
$$

for any $w \in \Sigma^*$. $\qquad\square$

In particular, considering the empty word, we infer from identity (4) that

$$
(||\mathcal{A}||, \varepsilon) = \sum_{q \in Q} \lambda_q \cdot \gamma_q = \lambda \cdot \gamma, \tag{5}
$$

as $\mu(\varepsilon) = E_Q$. Furthermore, identity (4) motivates the following definition.

3.3.7 Definition. Let Q be a non-empty finite set. A **linear representation** (of dimension Q over Σ and S) is a triple (λ, μ, γ) consisting of

- a row vector $\lambda \in S^Q$,

- a monoid homomorphism μ from Σ^* into $(S^{Q \times Q}, \cdot, E_Q)$ and

- a column vector $\gamma \in S^Q$.

It defines the series $r = ||(\lambda, \mu, \gamma)|| \in S\langle\!\langle \Sigma^* \rangle\!\rangle$ whose coefficients are determined by

$$
(r, w) = \lambda \cdot \mu(w) \cdot \gamma
$$

for $w \in \Sigma^*$.

The next result exhibits that the formalism of linear representations is expressively equivalent to the one of weighted automata.

3.3.8 Proposition. *Let $r \in S\langle\!\langle \Sigma^* \rangle\!\rangle$ be a formal power series. Then r is recognizable if and only if there exists a linear representation (λ, μ, γ) such that $r = ||(\lambda, \mu, \gamma)||$.*

Proof. If the power series $r \in S\langle\!\langle \Sigma^* \rangle\!\rangle$ is recognizable, then it is the behavior of some weighted automaton. Thus, Lemma 3.3.6 implies that $r = ||(\lambda, \mu, \gamma)||$ for some linear representation (λ, μ, γ).

For the converse, let $(\lambda', \mu', \gamma')$ be a linear representation of dimension Q. We define a weighted automaton $\mathcal{A} = (Q, \mathrm{in}, \mathrm{wt}, \mathrm{out})$ with state set Q by letting

$$\mathrm{in}(q) = \lambda'_q$$
$$\mathrm{wt}(p, a, q) = \mu'(a)_{p,q}$$
$$\mathrm{out}(q) = \gamma'_q$$

for $p, q \in Q$ and $a \in \Sigma$. Moreover, we consider the linear representation (λ, μ, γ) associated to $\mathcal{A}$ which is given by (1)–(3). If we establish the equality

$$(\lambda', \mu', \gamma') = (\lambda, \mu, \gamma),$$

then Lemma 3.3.6 implies $||\mathcal{A}|| = ||(\lambda', \mu', \gamma')||$. We have $\lambda' = \lambda$ and $\gamma' = \gamma$ by definition. For any letter $a \in \Sigma^*$ and states $p, q \in Q$, we obtain

$$\mu'(a)_{p,q} = \mathrm{wt}(p, a, q) = \mathrm{rweight}(paq) = \sum_{P:\, p \xrightarrow{a} q} \mathrm{rweight}(P) = \mu(a)_{p,q},$$

as $P = paq$ is the only path with label a leading from state p to state q. This shows that the restrictions $\mu'|_\Sigma$ and $\mu|_\Sigma$ coincide, and hence we obtain $\mu' = \mu$ by Lemma 2.1.12, which completes the proof. $\qquad\square$

This result explains why some authors, e.g. Berstel and Reutenauer [2, § I.5], use linear representations to define recognizable series or even weighted automata.

We further observe that – provided the operations of the semiring S are given in an effective way, i.e. S is *computable* – the proof of Proposition 3.3.8 provides effective procedures or programs P, P' such that:

- Provided with a weighted automaton $\mathcal{A}$, program P computes a linear representation (λ, μ, γ) that satisfies $||(\lambda, \mu, \gamma)|| = ||\mathcal{A}||$.

- Provided with a linear representation (λ, μ, γ), program P' computes a weighted automaton $\mathcal{A}$ that satisfies $||\mathcal{A}|| = ||(\lambda, \mu, \gamma)||$.

3.3.9 Example. Consider the alphabet $\Sigma = \{x\}$. In Example 3.2.12 we justified that the power series

$$r = \sum_{n \in \mathbb{N}} a_n x^n \in \mathbb{N}\langle\!\langle \Sigma^* \rangle\!\rangle$$

with $a_0 = a_1 = 1$ and $a_{n+2} = a_{n+1} + a_n$ for $n \in \mathbb{N}$ is recognizable. More precisely, we showed that the series r coincides with the behavior of the weighted automaton $\mathcal{A}^w$ associated to the following classical automaton:

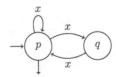

Following the proof of Proposition 3.3.8, we obtain a linear representation (λ, μ, γ) of dimension $Q = \{p, q\}$ such that $||(\lambda, \mu, \gamma)|| = r$ by setting

$$\lambda = \begin{pmatrix} \lambda_p & \lambda_q \end{pmatrix} = \begin{pmatrix} 1 & 0 \end{pmatrix},$$

$$\mu(x) = \begin{pmatrix} \mu(x)_{p,p} & \mu(x)_{p,q} \\ \mu(x)_{q,p} & \mu(x)_{q,q} \end{pmatrix} = \begin{pmatrix} 1 & 1 \\ 1 & 0 \end{pmatrix},$$

$$\gamma = \begin{pmatrix} \gamma_p \\ \gamma_q \end{pmatrix} = \begin{pmatrix} 1 \\ 0 \end{pmatrix}.$$

By Lemma 2.1.12, the map $\mu \colon \Sigma^* \to \mathbb{N}^{Q \times Q}$ is uniquely determined by its restriction to the alphabet $\Sigma = \{x\}$, i.e. by the matrix $\mu(x)$. More precisely, the image of any word x^n under μ is given by the matrix power

$$\mu(x^n) = \mu(x)^n = \begin{pmatrix} 1 & 1 \\ 1 & 0 \end{pmatrix}^n.$$

One can verify by induction that the (p,p)–entry of the matrix $\mu(x)^n$ is given by the Fibonacci number a_n, and thus we indeed have

$$\lambda \cdot \mu(x^n) \cdot \gamma = a_n = (r, x^n)$$

for any $n \in \mathbb{N}$.

A linear representation for the recognizable power series

$$\sum_{w \in \Sigma^*} |w|_a w$$

over the alphabet $\Sigma = \{a, b\}$, which we studied in Example 3.2.11, can be found in Berstel and Reutenauer [2, Chapter I, Example 5.3].

As we have already mentioned, employing linear representations instead of weighted automata makes it possible to use methods of Linear Algebra, and thus provides succinct algebraic proofs about recognizability. For instance, exploiting the equivalence in Proposition 3.3.8, we can now give a concise proof of Proposition 3.2.15:

Proof of Proposition 3.2.15. Let $h \colon \Sigma^* \to \Gamma^*$ be a monoid homomorphism and consider any power series $r \in S^{\mathrm{rec}}\langle\langle \Gamma^* \rangle\rangle$. Then there exists a linear representation (λ, μ, γ) (of dimension Q over Γ and S) such that $r = ||(\lambda, \mu, \gamma)||$. The composition

$$\mu \circ h \colon \Sigma^* \to S^{Q \times Q}$$

of the monoid homomorphisms $\mu \colon \Gamma^* \to S^{Q \times Q}$ and h is again a monoid homomorphism. Hence, $(\lambda, \mu \circ h, \gamma)$ is a linear representation (over Σ and S) fulfilling

$$(h^{-1}(r), w) = (r, h(w)) = \lambda \cdot \mu(h(w)) \cdot \gamma = \lambda \cdot (\mu \circ h)(w) \cdot \gamma = (||(\lambda, \mu \circ h, \gamma)||, w)$$

for any $w \in \Sigma^*$. Therefore, the series $h^{-1}(r) = ||(\lambda, \mu \circ h, \gamma)||$ belongs to $S^{\mathrm{rec}}\langle\langle \Sigma^* \rangle\rangle$. $\square$

The following result shows that not only the change of the underlying alphabet (see Proposition 3.2.15 and Proposition 3.2.16) but also the change of the underlying semiring preserves recognizability.

3.3.10 Proposition. *Let S, R be semirings and $h\colon S \to R$ a semiring homomorphism. The map*

$$S\langle\!\langle\Sigma^*\rangle\!\rangle \to R\langle\!\langle\Sigma^*\rangle\!\rangle, \quad r \mapsto h \circ r$$

preserves recognizability, i.e. if $r \in S^{\mathrm{rec}}\langle\!\langle\Sigma^*\rangle\!\rangle$, then the series

$$h \circ r = \sum_{w \in \Sigma^*} h((r, w))w$$

belongs to $R^{\mathrm{rec}}\langle\!\langle\Sigma^*\rangle\!\rangle$.

Proof. A detailed proof exploiting the characterization of recognizable power series by means of linear representations (see Proposition 3.3.8) can be found in Droste [11, Theorem 7.4]. $\square$

3.3.11 Example. Consider the alphabet $\Sigma = \{a, b\}$ and the series

$$|\cdot|_a = \sum_{w \in \Sigma^*} |w|_a w \in \mathbb{N}\langle\!\langle\Sigma^*\rangle\!\rangle.$$

In Example 3.2.11 we justified that this power series is recognizable.

a) It requires a simple computation to show that the map $h\colon \mathbb{N} \mapsto \mathbb{F}_2$ that assigns to each integer $n \in \mathbb{N}$ its residue modulo 2 constitutes a semiring homomorphism from the semiring $\mathbb{N}$ of natural numbers into the ring $\mathbb{F}_2$ of integers modulo 2. Thus, from the series $h \circ |\cdot|_a \in \mathbb{F}_2\langle\!\langle\Sigma^*\rangle\!\rangle$ is again recognizable by Proposition 3.3.10. Given $w \in \Sigma^*$, the corresponding coefficient $h(|w|_a)$ tells us whether the number of occurrences of the letter a in w is even or odd, more precisely

$$h(|w|_a) = \begin{cases} 0 & \text{if } |w|_a \text{ is even} \\ 1 & \text{if } |w|_a \text{ is odd.} \end{cases}$$

b) Similarly, one can verify that the map $h'\colon \mathbb{N} \to \mathbb{B}$ with $h(0) = 0$ and $h(\mathbb{N}^+) = \{1\}$ constitutes a semiring homomorphism from the semiring $\mathbb{N}$ of natural numbers into the Boolean semiring $\mathbb{B}$. Thus, the series $h' \circ |\cdot|_a \in \mathbb{B}\langle\!\langle\Sigma^*\rangle\!\rangle$ is again recognizable by Proposition 3.3.10. Given $w \in \Sigma^*$, the corresponding coefficient $h'(|w|_a)$ tells us whether the letter a does occur in w or not, more precisely

$$h'(|w|_a) = \begin{cases} 0 & \text{if } |w|_a = 0 \\ 1 & \text{if } |w|_a > 0. \end{cases}$$

For further investigations on the application of matrices in the context of (weighted) automata, we refer to Kuich and Salomaa [21], Salomaa and Soittola [35], Droste, Kuich and Vogler [8].

We have seen in Section 3.1 how formal power series generalize languages. In Section 3.2 we have introduced weighted automata, generalizing the classical ones, and thus we have naturally extended the notion of recognizability to formal power series. Furthermore, we have presented another very algebraic approach for the modeling of recognizable power series. The aim of the subsequent chapters is to also extend the classical characterizations of recognizability – in particular, the corresponding language-theoretic formalisms – from Chapter 2 to the realm of formal power series. Thus, we generalize the formalism of rationality in Chapter 4, and we prove the Kleene–Schützenberger Theorem, which is a quantitative extension of Kleene's Theorem. In Chapter 5, we introduce a weighted version of monadic second-order logic, which is due to Droste and Gastin [5], and derive their generalization of the Büchi–Elgot–Trakhtenbrot Theorem.

4. The Kleene–Schützenberger Theorem

Kleene [18] proved the equivalence of recognizability and rationality in the context of languages in 1956. In 1961, Schützenberger was the first to consider formal power series in the context of Automata Theory. Furthermore, in [37] he introduced the notion of weighted automata, which are quantitative extensions of the classical ones. In doing so, he extended the language-theoretic concept of recognizability to formal power series with coefficients in an arbitrary semiring. On the other hand, Schützenberger also investigated rational power series, which form a generalization of rational languages.

The main subject of this chapter is to derive Schützenberger's generalization of Kleene's classical result showing that the notions of recognizability and rationality are equivalent even in the realm of formal power series. In order to extend the language-theoretic concept of rationality to formal power series, in Section 4.1 we first define operations on the set $S\langle\langle \Sigma^* \rangle\rangle$ of formal power series that correspond to the language-theoretic operations union, intersection, concatenation, Kleene star and plus. With these operations, we introduce rational power series, which generalize the rational languages, and develop their fundamental theory in Section 4.2. In particular, we point out connections to the rational functions in one variable, as studied in Algebra. Section 4.3 is devoted to the proof of Schützenberger's remarkable result, referred to as Kleene–Schützenberger Theorem, which exhibits that the sets of recognizable and rational power series are identical. To derive that any rational power series belongs to the set $S^{\mathrm{rec}}\langle\langle \Sigma^* \rangle\rangle$ of recognizable power series, we prove that each of the operations defined in Section 4.1 preserves recognizability. In order to prove conversely that any recognizable power series is rational, we will exploit interconnections between weighted automata, rational formal power series and linear systems of equations. We conclude this chapter by some remarks on decidability properties and applications of rational and recognizable power series, respectively.

We recall that throughout this work Σ denotes an alphabet and S denotes a non-trivial semiring.

4.1. Operations on Formal Power Series

The aim of this section is to extend all language-theoretic notions in the context of rationality to the realm of formal power series. More precisely, we first define a counterpart of singleton languages in the realm of formal power series. Then we introduce operations on the set $S\langle\langle \Sigma^* \rangle\rangle$ of formal power series that arise as natural generalizations of the language-theoretic rational operations union, concatenation and star. Equipped with these operations, the set $S\langle\langle\Sigma^*\rangle\rangle$ of formal power series even constitutes a semiring, generalizing the semiring of formal languages (see Proposition 4.1.7). We study two further operations on the set $S\langle\langle\Sigma^*\rangle\rangle$, one of which generalizes the intersection of languages and the other does not have a natural counterpart in Language Theory. The primary source for this section is Droste and Kuske [12, § 4]. However, we also took inspiration from Droste and Kuich [9, § 3] and Droste [11, § 6].

We start by generalizing singleton languages, which are the basic components of rational languages.

4.1.1 Definition.

a) A formal power series $r \in S\langle\langle\Sigma^*\rangle\rangle$ is called **monomial** if its support contains at most one word, i.e. if $|\operatorname{supp}(r)| \leq 1$. Given a word $w \in \Sigma^*$ and an element $s \in S$, we write sw for the monomial in $S\langle\langle\Sigma^*\rangle\rangle$ defined by

$$(sw, w) := s \text{ and } (sw, v) := 0 \text{ for any } w \neq v \in \Sigma^*.$$

b) We denote by

$$S\langle\Sigma^*\rangle := \{r \in S\langle\langle\Sigma^*\rangle\rangle \mid |\operatorname{supp}(r)| < \infty\}$$

the set of all formal power series in $S\langle\langle\Sigma^*\rangle\rangle$ with finite support. The power series in $S\langle\Sigma^*\rangle$ are called **polynomials**.

Clearly, any monomial $m \in S\langle\langle\Sigma^*\rangle\rangle$ is of the form $m = sw$ for some $w \in \Sigma^*$ and $s \in S$. If $s = 0$, then the monomial $0w$ coincides with the constant series $\mathbb{1}_\emptyset$, i.e. we have

$$0w = \mathbb{1}_\emptyset = \sum_{v \in \Sigma^*} 0v.$$

In this case, the support of $0w$ is given by the empty language

$$\operatorname{supp}(0w) = \operatorname{supp}(\mathbb{1}_\emptyset) = \emptyset.$$

On the other hand, if $s \neq 0$, then the support of the monomial sw is given by the singleton language

$$\operatorname{supp}(sw) = \{w\}.$$

Thus, monomials are indeed the counterpart of singleton languages in the realm of formal power series. Similarly, polynomials generalize finite languages.

Next, we extend the language-theoretic operations union, intersection, concatenation, star and plus to formal power series.

4.1.2 Definition. Given $r_1, r_2 \in S\langle\langle\Sigma^*\rangle\rangle$, their **sum** $r_1 + r_2 \in S\langle\langle\Sigma^*\rangle\rangle$ is defined pointwise by

$$(r_1 + r_2, w) := (r_1, w) + (r_2, w)$$

for any $w \in \Sigma^*$, i.e. we have

$$r_1 + r_2 = \left(\sum_{w \in \Sigma^*} (r_1, w)w\right) + \left(\sum_{w \in \Sigma^*} (r_2, w)w\right) = \sum_{w \in \Sigma^*} ((r_1, w) + (r_2, w))w.$$

This operation is associative, commutative, and as neutral element it has the constant series $\mathbb{1}_\emptyset$ with value $(\mathbb{1}_\emptyset, w) = 0$ for any $w \in \Sigma^*$. Thus, $(S\langle\!\langle\Sigma^*\rangle\!\rangle, +, \mathbb{1}_\emptyset)$ forms a commutative monoid. Furthermore, the pointwise sum of formal power series generalizes the union of languages. More precisely, for the Boolean semiring $\mathbb{B}$ we obtain the following correspondence:

$$\operatorname{supp}(r_1 + r_2) = \operatorname{supp}(r_1) \cup \operatorname{supp}(r_2)$$
$$\mathbb{1}_{L_1 \cup L_2} = \mathbb{1}_{L_1} + \mathbb{1}_{L_2} \tag{1}$$

where L_1, L_2 are languages over Σ and $r_1, r_2 \in \mathbb{B}\langle\!\langle\Sigma^*\rangle\!\rangle$ are formal power series. Crucial for this correspondence is that a sum in $\mathbb{B}$ equals 1 if and only if at least one summand has value 1, and otherwise the sum equals 0. Indeed, this yields

$$w \in \operatorname{supp}(r_1 + r_2) \iff (r_1 + r_2, w) = (r_1, w) + (r_2, w) = 1$$
$$\iff (r_1, w) = 1 \text{ or } (r_2, w) = 1$$
$$\iff w \in \operatorname{supp}(r_1) \cup \operatorname{supp}(r_2)$$

for any $w \in \Sigma^*$.

As any family of languages has a union, one is tempted to also define the sum of arbitrarily many formal power series. This fails in general, since it would require the sum of infinitely many elements of the semiring S (which, e.g. in $S = \mathbb{N}$, does not exist). However, certain families can be summed.

4.1.3 Definition. Let $\{r_i\}_{i \in I}$ be a family of formal power series in $S\langle\!\langle\Sigma^*\rangle\!\rangle$, where I is an arbitrary index set.

a) We call the family $\{r_i\}_{i \in I}$ **locally finite** if for any word $w \in \Sigma^*$ there are only finitely many $i \in I$ with $(r_i, w) \neq 0$.

b) If the family $\{r_i\}_{i \in I}$ is locally finite, then the **sum** $\sum_{i \in I} r_i \in S\langle\!\langle\Sigma^*\rangle\!\rangle$ is defined by

$$\left(\sum_{i \in I} r_i, w \right) := \underbrace{\sum_{\substack{i \in I \\ (r_i, w) \neq 0}} (r_i, w)}_{\text{finite sum in } S}$$

for any $w \in \Sigma^*$, i.e. we have

$$\sum_{i \in I} r_i = \sum_{i \in I} \left(\sum_{w \in \Sigma^*} (r_i, w)w \right) = \sum_{w \in \Sigma^*} \left(\sum_{\substack{i \in I \\ (r_i, w) \neq 0}} (r_i, w) \right) w.$$

If $I = \{1, \ldots, n\}$ is a finite set with $n \in \mathbb{N}$, then clearly any family of formal power series $\{r_i\}_{i \in I}$ in $S\langle\!\langle\Sigma^*\rangle\!\rangle$ is locally finite and we obtain

$$\sum_{i \in I} r_i = \sum_{i=1}^{n} r_i = \sum_{w \in \Sigma^*} \left(\sum_{i=1}^{n} (r_i, w) \right) w.$$

Given a formal power series $r \in S\langle\!\langle \Sigma^* \rangle\!\rangle$, the family $\{(r,w)w\}_{w \in \Sigma^*}$ of monomials and its subfamily $\{(r,w)w\}_{w \in \mathrm{supp}(r)}$ are both locally finite, since we have $\mathrm{supp}((r,w)w) \subseteq \{w\}$ for any $w \in \Sigma^*$. Therefore, we can sum the families to obtain

$$r = \sum_{w \in \Sigma^*} (r,w)w = \sum_{w \in \mathrm{supp}(r)} (r,w)w,$$

which perfectly suits the usual notation of power series as formal sums. If r is a polynomial, then the family $\{(r,w)w\}_{w \in \mathrm{supp}(r)}$ is finite. In particular, any polynomial is a finite sum of polynomials, as it is the case for classical polynomials. We have mentioned in Section 3.1 that in the literature the elements of $S\langle\!\langle \Sigma^* \rangle\!\rangle$ are often called *formal power series with non-commuting variables*. Since the monomials in $S\langle\!\langle \Sigma^* \rangle\!\rangle$ are of the form sw with $w \in \Sigma^*$ and $s \in S$, we may consider the words in Σ^* as the variables of power series in $S\langle\!\langle \Sigma^* \rangle\!\rangle$. As a result, the variables are *non-commuting*, since Σ^* is a non-commutative monoid (if $|\Sigma| \geq 2$).

4.1.4 Definition. Given $r_1, r_2 \in S\langle\!\langle \Sigma^* \rangle\!\rangle$, their **Hadamard product** $r_1 \odot r_2 \in S\langle\!\langle \Sigma^* \rangle\!\rangle$ is defined pointwise by

$$(r_1 \odot r_2, w) := (r_1, w) \cdot (r_2, w)$$

for any $w \in \Sigma^*$, i.e. we have

$$\left(\sum_{w \in \Sigma^*} (r_1, w)w \right) \odot \left(\sum_{w \in \Sigma^*} (r_2, w)w \right) = \sum_{w \in \Sigma^*} ((r_1, w) \cdot (r_2, w))w.$$

This operation is associative, distributes over addition from both sides, and has the constant power series $\mathbb{1}_{\Sigma^*}$, with value $(\mathbb{1}_{\Sigma^*}, w) = 1$ for any $w \in \Sigma^*$, as neutral element. Thus, $(S\langle\!\langle \Sigma^* \rangle\!\rangle, +, \odot, \mathbb{1}_\emptyset, \mathbb{1}_{\Sigma^*})$ constitutes a semiring, which is commutative if and only if the underlying semiring S is commutative. Furthermore, the Hadamard product of formal power series is the counterpart of the intersection of languages. More precisely, for the Boolean semiring $\mathbb{B}$ we obtain the following correspondence:

$$\mathrm{supp}(r_1 \odot r_2) = \mathrm{supp}(r_1) \cap \mathrm{supp}(r_2)$$
$$\mathbb{1}_{L_1 \cap L_2} = \mathbb{1}_{L_1} \odot \mathbb{1}_{L_2} \tag{2}$$

where L_1, L_2 are languages over Σ and $r_1, r_2 \in \mathbb{B}\langle\!\langle \Sigma^* \rangle\!\rangle$ are formal power series. Crucial for this correspondence is that a product in $\mathbb{B}$ equals 1 if and only if all factors have value 1, and otherwise the product equals 0. Indeed, this yields

$$w \in \mathrm{supp}(r_1 \odot r_2) \iff (r_1 \odot r_2, w) = (r_1, w) \cdot (r_2, w) = 1$$
$$\iff (r_1, w) = 1 \text{ and } (r_2, w) = 1$$
$$\iff w \in \mathrm{supp}(r_1) \cap \mathrm{supp}(r_2)$$

for any $w \in \Sigma^*$.

4.1.5 Remark. Let S be a ring and consider the alphabet $\Sigma = \{x\}$. We have already seen in Remark 3.1.11 that the set $S\langle\!\langle \{x\}^* \rangle\!\rangle$ coincides precisely with the set $S[[x]]$ of formal power series over S in one variable, since any formal power series in $S\langle\langle\{x\}^*\rangle\rangle$ can be written as

$$\sum_{n\in\mathbb{N}} a_n x^n$$

with $a_n \in S$ for any $n \in \mathbb{N}$. Similarly, the set $S\langle\{x\}^*\rangle$ coincides with the set $S[x]$ of polynomials over S in one variable. With their usual addition and multiplication, the sets $S[[x]]$ and $S[x]$ both constitute rings. However, the usual product of power series and polynomials in $S[[x]]$ and $S[x]$, respectively, is defined by some kind of a (discrete) convolution, whereas the Hadamard product is defined by pointwise multiplication of the coefficients. Therefore, we introduce another fundamental multiplicative operation on power series in $S\langle\langle\Sigma^*\rangle\rangle$, generalizing the one in $S[[x]]$ and $S[x]$.

4.1.6 Definition. Given $r_1, r_2 \in S\langle\langle\Sigma^*\rangle\rangle$, their **Cauchy product** $r_1 \cdot r_2 \in S\langle\langle\Sigma^*\rangle\rangle$ is defined by

$$(r_1 \cdot r_2, w) := \sum_{\substack{u,v\in\Sigma^* \\ w=uv}} (r_1, u) \cdot (r_2, v)$$

for any $w \in \Sigma^*$, i.e. we have

$$\left(\sum_{w\in\Sigma^*} (r_1, w)w\right) \cdot \left(\sum_{w\in\Sigma^*} (r_2, w)w\right) = \sum_{w\in\Sigma^*} \left(\sum_{\substack{u,v\in\Sigma^* \\ w=uv}} (r_1, u) \cdot (r_2, v)\right) w.$$

Any word $w \in \Sigma^*$ has only finitely many factorizations $w = uv$ with $u, v \in \Sigma^*$, implying that also the sum

$$(r_1 \cdot r_2, w) = \sum_{\substack{u,v\in\Sigma^* \\ w=uv}} (r_1, u) \cdot (r_2, v)$$

is finite. Hence, the Cauchy product $r_1 \cdot r_2$ is well-defined. The Cauchy product is an associative operation and has the monomial $1\varepsilon = \mathbb{1}_{\{\varepsilon\}}$ as neutral element. Indeed, exploiting the distributivity of S, one can easily verify the first statement. For the latter, we consider any power series $r \in S\langle\langle\Sigma^*\rangle\rangle$ and any word $w \in \Sigma^*$ to obtain

$$(1\varepsilon \cdot r, w) = \sum_{w=uv} \underbrace{(1\varepsilon, u)}_{\substack{=0 \\ \text{if } u\neq\varepsilon}} \cdot (r, v) = \underbrace{(1\varepsilon, \varepsilon)}_{=1} \cdot (r, w) = (r, w),$$

which means $1\varepsilon \cdot r = r$. Similarly, one can show that also $r \cdot 1\varepsilon = r$ for any $r \in S\langle\langle\Sigma^*\rangle\rangle$. Now let I be an arbitrary index set and consider a locally finite family $\{r_i\}_{i\in I}$ of formal power series in $S\langle\langle\Sigma^*\rangle\rangle$. For each $r \in S\langle\langle\Sigma^*\rangle\rangle$, the families $\{r \cdot r_i\}_{i\in I}$ and $\{r_i \cdot r\}_{i\in I}$ are again locally finite, as by definition 0 annihilates the whole semiring S. Moreover, it requires a short computation to show that we have

$$r \cdot \sum_{i\in I} r_i = \sum_{i\in I} r \cdot r_i \quad \text{and} \quad \left(\sum_{i\in I} r_i\right) \cdot r = \sum_{i\in I} r_i \cdot r$$

(cf. Droste and Kuich [9, page 13]).

We infer from this that the Cauchy product distributes over addition from both sides, and hence the tuple

$$(S\langle\!\langle \Sigma^* \rangle\!\rangle, +, \cdot, \mathbb{1}_\emptyset, 1\varepsilon)$$

constitutes a semiring, called the **semiring of formal power series** over Σ and S. Furthermore, the Cauchy product extends the language-theoretic concatenation to the realm of formal power series. More precisely, for the Boolean semiring $\mathbb{B}$ we obtain the following correspondence:

$$\text{supp}(r_1 \cdot r_2) = \text{supp}(r_1) \cdot \text{supp}(r_2)$$
$$\mathbb{1}_{L_1 \cdot L_2} = \mathbb{1}_{L_1} \cdot \mathbb{1}_{L_2}$$

$$(3)$$

where L_1, L_2 are languages over Σ and $r_1, r_2 \in \mathbb{B}\langle\!\langle \Sigma^* \rangle\!\rangle$ are formal power series. Indeed, exploiting the arithmetic of the Boolean semiring yields

$$w \in \text{supp}(r_1 \cdot r_2) \;\Leftrightarrow\; (r_1 \cdot r_2, w) = \sum_{\substack{u,v \in \Sigma^* \\ w=uv}} (r_1, u) \cdot (r_2, v) = 1$$

$$\Leftrightarrow (r_1, u) \cdot (r_2, v) = 1 \text{ for some } u, v \in \Sigma^* \text{ with } uv = w$$
$$\Leftrightarrow (r_1, u) = 1 \text{ and } (r_2, v) = 1 \text{ for some } u, v \in \Sigma^* \text{ with } uv = w$$
$$\Leftrightarrow u \in \text{supp}(r_1) \text{ and } v \in \text{supp}(r_2) \text{ for some } u, v \in \Sigma^* \text{ with } uv = w$$
$$\Leftrightarrow w \in \text{supp}(r_1) \cdot \text{supp}(r_2)$$

for any $w \in \Sigma^*$.

Considering still the Boolean semiring $\mathbb{B}$, the correspondences (1), (2), (3) and the identity $1\varepsilon = \mathbb{1}_{\{\varepsilon\}}$ yield the following isomorphism result.

4.1.7 Proposition. *The semirings*

$$(\mathbb{B}\langle\!\langle \Sigma^* \rangle\!\rangle, +, \odot, \mathbb{1}_\emptyset, \mathbb{1}_{\Sigma^*}) \quad and \quad (\mathcal{P}(\Sigma^*), \cup, \cap, \emptyset, \Sigma^*),$$

as well as $\quad (\mathbb{B}\langle\!\langle \Sigma^* \rangle\!\rangle, +, \cdot, \mathbb{1}_\emptyset, 1\varepsilon) \quad and \quad (\mathcal{P}(\Sigma^*), \cup, \cdot, \emptyset, \{\varepsilon\})$

are isomorphic via the map

$$\text{supp}\colon \mathbb{B}\langle\!\langle \Sigma^* \rangle\!\rangle \to \mathcal{P}(\Sigma^*), \; r \mapsto \text{supp}(r)$$

and its inverse

$$\text{char}\colon \mathcal{P}(\Sigma^*) \to \mathbb{B}\langle\!\langle \Sigma^* \rangle\!\rangle, \; L \mapsto \mathbb{1}_L.$$

When we refer to the semiring $S\langle\!\langle \Sigma^* \rangle\!\rangle$ (of formal power series) in the following, then we mean the semiring $(S\langle\!\langle \Sigma^* \rangle\!\rangle, +, \cdot, \mathbb{1}_\emptyset, 1\varepsilon)$ equipped with the Cauchy product (and not with the Hadamard product).

In Proposition 3.2.15 and Proposition 3.2.16 we have exhibited that the transformations from Definition 3.2.13 preserve recognizability. The next results, which can be found in Droste and Kuich [9, Proposition 3.6], Kuich and Salomaa [21, Chapter I, Theorem 6.1], show that these transformations are even compatible with sum and Cauchy product of formal power series.

4.1.8 Proposition. *Let Σ, Γ be alphabets and $h\colon \Sigma^* \to \Gamma^*$ a length-preserving monoid homomorphism. The map*

$$h^{-1}\colon S\langle\!\langle\Gamma^*\rangle\!\rangle \to S\langle\!\langle\Sigma^*\rangle\!\rangle,$$

which is defined by

$$h^{-1}(r) = r \circ h = \sum_{w \in \Sigma^*} (r, h(w))w$$

for any $r \in S^{\mathrm{rec}}\langle\!\langle\Gamma^*\rangle\!\rangle$, constitutes a semiring homomorphism from the semiring $S\langle\!\langle\Gamma^*\rangle\!\rangle$ into the semiring $S\langle\!\langle\Sigma^*\rangle\!\rangle$.

4.1.9 Proposition. *Let Σ, Γ be alphabets and $h\colon \Sigma^* \to \Gamma^*$ a non-deleting monoid homomorphism. The map*

$$h\colon S\langle\!\langle\Sigma^*\rangle\!\rangle \to S\langle\!\langle\Gamma^*\rangle\!\rangle,$$

which is defined by

$$h(r) = \sum_{v \in \Gamma^*} \left(\sum_{w \in h^{-1}(v)} (r, w) \right) v$$

for any $r \in S^{\mathrm{rec}}\langle\!\langle\Sigma^*\rangle\!\rangle$, constitutes a semiring homomorphism from the semiring $S\langle\!\langle\Sigma^*\rangle\!\rangle$ into the semiring $S\langle\!\langle\Gamma^*\rangle\!\rangle$.

We still have to extend a fundamental language-theoretic operation: the Kleene star. As we have already found the counterpart of the language-theoretic concatenation, we can naturally define powers of a formal power series.

4.1.10 Definition. Let $r \in S\langle\!\langle\Sigma^*\rangle\!\rangle$ be a formal power series. We define **powers** of r inductively by

$$r^0 := 1\varepsilon,$$
$$r^{n+1} := r^n \cdot r \quad (\text{for } n \in \mathbb{N}).$$

Considering e.g. $n = 3$ and exploiting the distributivity in S, we obtain

$$(r^3, w) = \sum_{w=uv} (r^2, u) \cdot (r, v)$$
$$= \sum_{w=uv} \left(\sum_{v=xy} (r, x) \cdot (r, y) \right) \cdot (r, v)$$
$$= \sum_{w=uv} \sum_{v=xy} (r, x) \cdot (r, y) \cdot (r, v)$$
$$= \sum_{w=xyv} (r, x) \cdot (r, y) \cdot (r, v).$$

One can proceed inductively to prove that for any $n \in \mathbb{N}$ and any non-empty $w \in \Sigma^+$ we have

$$(r^n, w) = \sum_{w=w_1\ldots w_n} \prod_{i=1}^{n} (r, w_i) = \sum_{w=w_1\ldots w_n} (r, w_1) \cdot \ldots \cdot (r, w_n).$$

Considering the empty word, we obtain $(r^n, \varepsilon) = (r, \varepsilon)^n$ for any $n \in \mathbb{N}$. Furthermore, we immediately infer from the correspondence (3) that for the Boolean semiring $\mathbb{B}$ we have

$$\mathrm{supp}(r^n) = (\mathrm{supp}(r))^n$$
$$\mathbb{1}_{L^n} = (\mathbb{1}_L)^n$$

where $n \in \mathbb{N}$, L is a language over Σ and $r \in \mathbb{B}\langle\!\langle \Sigma^* \rangle\!\rangle$ is a formal power series.

Next, we want to the define the counterpart of the language-theoretic Kleene star. For a language L, its Kleene star L^* is defined to be the union of all the powers L^n of L, where $n \in \mathbb{N}$. To also define the Kleene star r^* of a formal power series r, one would therefore try to sum all powers r^n of r, where $n \in \mathbb{N}$. However, the family $\{r^n\}_{n \in \mathbb{N}}$ is in general not locally finite and hence cannot be summed. For instance, considering the semiring $\mathbb{N}$ of natural numbers and the monomial 1ε, we obtain

$$((1\varepsilon)^n, \varepsilon) = (1\varepsilon, \varepsilon) = 1 \neq 0$$

for any $n \in \mathbb{N}$. Therefore, we need to restrict the Kleene star in the context of formal power series.

4.1.11 Definition. Let $r \in S\langle\!\langle \Sigma^* \rangle\!\rangle$ be a formal power series.

a) We call r **proper** if it fulfills $(r, \varepsilon) = 0$, i.e. if the *constant term* of r vanishes.

b) If r is proper, then its **(Kleene) star** $r^* \in S\langle\!\langle \Sigma^* \rangle\!\rangle$ is defined by

$$r^* := \sum_{n \in \mathbb{N}} r^n.$$

c) If r is proper, then its **(Kleene) plus** $r^+ \in S\langle\!\langle \Sigma^* \rangle\!\rangle$ is defined by

$$r^+ := \sum_{n \in \mathbb{N}^+} r^n.$$

4.1.12 Remark. In the study of generalized power series fields, a sum of the form

$$\sum_{n \in \mathbb{N}} r^n,$$

where r is a generalized power series whose support only consists of positive elements, is again a generalized power series (cf. Neumann's Lemma [23, page 210 f.]) satisfying the identity

$$\sum_{n \in \mathbb{N}} r^n = (1 - r)^{-1}$$

(cf. Krapp, Kuhlmann and Serra [19, page 945 f.] as well as Kuhlmann [20, Script 11]). Hence, from an algebraic point of view, the Kleene star r^* of a power series r can also be referred to as **Neumann sum** of r. However, since the literature uses the term *Kleene star* both for languages and power series, we also follow this terminology.

Given a proper formal power series r, a word w and an $n > |w|$, we have

$$(r^n, w) = \sum_{w=w_1 \ldots w_n} (r, w_1) \cdot \ldots \cdot (r, w_n) = 0,$$

since $w = w_1 \ldots w_n$ implies $w_i = \varepsilon$ for some $i \in \{1, \ldots, n\}$ and thus $(r, w_i) = (r, \varepsilon) = 0$ by assumption. Hence, the families $\{r^n\}_{n \in \mathbb{N}}$ and $\{r^n\}_{n \in \mathbb{N}^+}$ of powers are both locally finite and therefore can be summed. In particular, the star and the plus are well-defined for proper formal power series. More precisely, given a proper formal power series $r \in S\langle\!\langle \Sigma^* \rangle\!\rangle$ and a non-empty word $w \in \Sigma^+$, we compute

$$(r^*, w) = \left(\sum_{n \in \mathbb{N}} r^n, w \right) = \sum_{n=0}^{|w|} (r^n, w)$$

$$= \sum_{n=0}^{|w|} \sum_{\substack{w=w_1 \ldots w_n \\ w_i \in \Sigma^*}} \prod_{i=1}^{n} \underbrace{(r, w_i)}_{\substack{=0 \\ \text{if } w_i = \varepsilon}}$$

Thus, we obtain

$$(r^*, w) = \sum_{\substack{w=w_1 \ldots w_k \\ k \in \mathbb{N}^+, w_i \in \Sigma^+}} \prod_{i=1}^{k} (r, w_i)$$

for any non-empty $w \in \Sigma^+$. Considering the empty word, we obtain

$$(r^*, \varepsilon) = (r^0, \varepsilon) = (1\varepsilon, \varepsilon) = 1.$$

We observe further that any proper formal power series $r \in S\langle\!\langle \Sigma^* \rangle\!\rangle$ fulfills the equations

$$r^* = 1\varepsilon + r^+,$$

$$r^+ = r \cdot r^*.$$

If S is a ring, then so is $S\langle\!\langle \Sigma^* \rangle\!\rangle$ (cf. Salomaa and Soittola [35, page 12]) and in this case the star r^* is the multiplicative inverse of the series $1\varepsilon - r$, which yields the classical identity

$$r^* = \sum_{n \in \mathbb{N}} r^n = (1 - r)^{-1}. \qquad \text{(Neumann sum)}$$

Indeed, 1ε is the neutral element of multiplication in $S\langle\!\langle \Sigma^* \rangle\!\rangle$ and the equations above imply

$$(1\varepsilon - r) \cdot r^* = r^* - r \cdot r^* = r^* - r^+ = 1\varepsilon.$$

In the literature, proper power series are often referred to as *quasi-regular* power series and the plus r^+ of such a power series is referred to as *quasi-inverse* of r (cf. Kuich and Salomaa [21, Chapter 1], Salomaa and Soittola [35, § I.3]). The notion of proper power series is generalized by the one of *cycle-free* power series (cf. Droste and Kuich [9, § 3]).

In [21, § I.2 and § I.3] and [9, § 3 and § 5] one finds several identities for proper and cycle-free power series, respectively. Moreover, it can be shown that the semiring homomorphisms in Proposition 4.1.8 and Proposition 4.1.9 are also compatible with the star of formal power series (cf. [9, Proposition 3.6]).

The Kleene star of formal power series generalizes the language-theoretic Kleene star. More precisely, for the Boolean semiring $\mathbb{B}$ we obtain the following correspondence:

$$\mathrm{supp}(r^*) = (\mathrm{supp}(r))^* \tag{4}$$
$$\mathbb{1}_{L^*} = (\mathbb{1}_L)^*$$

where $L \subseteq \Sigma^+$ is a language over Σ not containing the empty word and $r \in \mathbb{B}\langle\!\langle \Sigma^* \rangle\!\rangle$ is a proper formal power series. Crucial for the proof of this correspondence is that a power series is proper if and only if its support does not contain the empty word. Indeed, for any non-empty word $w \in \Sigma^+$ the coefficient (r^*, w) is given by

$$(r^*, w) = \sum_{\substack{w = w_1 \dots w_k \\ k \in \mathbb{N}^+, w_i \in \Sigma^+}} \prod_{i=1}^{k} (r, w_i).$$

Thus, exploiting the arithmetic of the Boolean semiring yields

$$w \in \mathrm{supp}(r^*) \ \Leftrightarrow \ (r^*, w) = 1$$

$$\Leftrightarrow \ \sum_{\substack{w = w_1 \dots w_k \\ k \in \mathbb{N}^+, w_i \in \Sigma^+}} \prod_{i=1}^{k} (r, w_i) = 1$$

$$\Leftrightarrow \text{ there exist } k \in \mathbb{N}^+ \text{ and } w_1, \dots, w_k \in \Sigma^+ \text{ such that}$$
$$w = w_1 \dots w_k \text{ and } (r, w_i) = 1 \text{ for any } i \in \{1, \dots, k\}$$

$$\Leftrightarrow \text{ there exist } k \in \mathbb{N}^+ \text{ and } w_1, \dots, w_k \in \mathrm{supp}(r) \text{ such that}$$
$$w = w_1 \dots w_k$$

$$\Leftrightarrow \ w \in (\mathrm{supp}(r))^*$$

for any non-empty $w \in \Sigma^+$, as $\mathrm{supp}(r) \subseteq \Sigma^+$. The empty word belongs both to $\mathrm{supp}(r^*)$ and $(\mathrm{supp}(r))^*$, as $(r^*, \varepsilon) = (r^0, \varepsilon) = 1$ and $\varepsilon \in (\mathrm{supp}(r))^0 \subseteq (\mathrm{supp}(r))^*$.

We introduce another simple operation, which does not have a natural language-theoretic counterpart, but which is very useful when working with formal power series.

4.1.13 Definition. Let $r \in S\langle\!\langle \Sigma^* \rangle\!\rangle$ be a formal power series and let $s \in S$. The **(left) scalar product** $s \cdot r \in S\langle\!\langle \Sigma^* \rangle\!\rangle$ is defined by

$$(s \cdot r, w) := s \cdot (r, w)$$

for any $w \in \Sigma^*$, i.e. we have

$$s \cdot \sum_{w \in \Sigma^*} (r, w) w = \sum_{w \in \Sigma^*} (s \cdot (r, w)) w.$$

Sometimes $s \cdot r$ is simply written sr.

Each monomial in $S\langle\!\langle \Sigma^* \rangle\!\rangle$ can be written as a scalar product. More precisely, given a word $w \in \Sigma^*$ and $s \in S$, we have $1w = \mathbb{1}_{\{w\}}$ and $sw = s \cdot 1w = s \cdot \mathbb{1}_{\{w\}}$. Furthermore, we can express each scalar product as a Hadamard product and as a Cauchy product:

$$sr = \underline{s} \odot r,$$

$$sr = s\varepsilon \cdot r,$$

where $r \in S\langle\!\langle \Sigma^* \rangle\!\rangle$, $s \in S$ and $\underline{s}$ denotes the constant power series

$$\underline{s} = \sum_{w \in \Sigma^*} sw = s \cdot \mathbb{1}_{\Sigma^*}.$$

Having extended the rational language-theoretic operations union, concatenation and star to the realm of formal power series, we can now also generalize the concept of rationality.

4.2. Rational Formal Power Series

The aim of this section is to develop the fundamental theory of rational formal power series, which originally was initiated by Schützenberger [37] in 1961. In particular, we describe how it naturally extends the language-theoretic concept of rationality from Section 2.3. In Remark 4.2.7 we point out a connection between rational power series and rational functions in one variable, as studied in Algebra.

We mainly follow Droste and Kuske [12, § 4]. However, we also took inspiration from Berstel and Reutenauer [2, Chapter I] and Droste [11, § 6].

4.2.1 Definition. We define the subset $S^{\mathrm{rat}}\langle\!\langle \Sigma^* \rangle\!\rangle \subseteq S\langle\!\langle \Sigma^* \rangle\!\rangle$ of **rational formal power series** inductively, as follows:

(1) For each $s \in S$, the monomial $s\varepsilon$ is a rational formal power series.

(2) For each $a \in \Sigma$, the monomial $1a$ is a rational formal power series.

(3) If r_1 and r_2 are rational formal power series, then so is their sum $r_1 + r_2$.

(4) If r_1 and r_2 are rational formal power series, then so is their Cauchy product $r_1 \cdot r_2$.

(5) If r is proper and a rational formal power series, then its Kleene star r^* is again a rational formal power series.

Sum, Cauchy product and Kleene star are called **rational operations**.

Rational power series are often specified using so-called *weighted rational expressions*, generalizing the rational expressions which are used to describe rational languages (cf. Sakarovitch [34, page 120 f.]).

4.2.2 Remark. In Chapter 2 we have inferred directly from the definition of rational languages that every finite language is rational (see Remark 2.3.2). Extending this to formal power series, we infer now from Definition 4.2.1 that every polynomial is rational:

a) The monomial 1ε is rational by step (1) from Definition 4.2.1. Steps (2) and (4) imply that the Cauchy product

$$1w = 1a_1 \cdot \ldots \cdot 1a_n$$

is rational for any non-empty word $w = a_1 \ldots a_n \in \Sigma^+$. Therefore, any monomial of the form $1w$ with $w \in \Sigma^*$ belongs to $S^{\mathrm{rat}}\langle\!\langle \Sigma^* \rangle\!\rangle$.

b) Scalar multiplication preserves rationality. Indeed, given a rational formal power series $r \in S^{\mathrm{rat}}\langle\!\langle \Sigma^* \rangle\!\rangle$ and an $s \in S$, the scalar product sr can be written as Cauchy product $sr = s\varepsilon \cdot r$ and is thus rational by steps (1) and (4) from Definition 4.2.1.

c) Every polynomial in $S\langle\!\langle \Sigma^* \rangle\!\rangle$ is rational, i.e. we have $S\langle \Sigma^* \rangle \subseteq S^{\mathrm{rat}}\langle\!\langle \Sigma^* \rangle\!\rangle$. Indeed, any monomial sw with $w \in \Sigma^*$ and $s \in S$ can be written as scalar product $sw = s \cdot 1w$ and is thus rational by a) and b). Hence, any polynomial, i.e. any finite sum of monomials, is rational by step (2) from Definition 4.2.1.

In particular, a power series is rational if and only if it can be constructed from finitely many polynomials using the rational operations sum, Cauchy product and Kleene star, where the Kleene star is only applied to proper series. Therefore, $S^{\mathrm{rat}}\langle\!\langle \Sigma^* \rangle\!\rangle$ is the smallest subset of $S\langle\!\langle \Sigma^* \rangle\!\rangle$ that contains the set $S\langle \Sigma^* \rangle$ of polynomials and is closed under sum, Cauchy product and Kleene star (applied to proper series only).

The following result shows that for the Boolean semiring $\mathbb{B}$ the notion of rational power series coincides with the one of rational languages, and hence it may be regarded as a counterpart of Proposition 3.2.7 in the context of rationality.

4.2.3 Proposition. *The map*

$$\mathrm{char}\colon \mathcal{P}(\Sigma^*) \to \mathbb{B}\langle\!\langle \Sigma^* \rangle\!\rangle, \quad L \mapsto \mathbb{1}_L$$

and its inverse

$$\mathrm{supp}\colon \mathbb{B}\langle\!\langle \Sigma^* \rangle\!\rangle \to \mathcal{P}(\Sigma^*), \quad r \mapsto \mathrm{supp}(r)$$

provide a one-to-one correspondence between the rational languages over Σ and the set $\mathbb{B}^{\mathrm{rat}}\langle\!\langle \Sigma^* \rangle\!\rangle$ of rational power series from Σ^* into the Boolean semiring $\mathbb{B}$.

Proof. First, we observe that the formal power series obtained by steps (1) and (2) from Definition 4.2.1 are all monomials. In particular, their supports are either empty or singletons and thus rational languages. Furthermore, the correspondences (1), (3) and (4) from Section 4.1 show that the map supp is compatible with the rational operations on formal power series and languages, respectively. Hence, given a rational power series $r \in \mathbb{B}^{\mathrm{rat}}\langle\!\langle \Sigma^* \rangle\!\rangle$, we infer that $\mathrm{supp}(r)$ is a rational language.

For the converse, let L be an arbitrary rational language over Σ. We want to derive that its characteristic series $\mathbb{1}_L$ is again rational. However, the language-theoretic Kleene star is unrestricted, whereas the Kleene star in $\mathbb{B}\langle\!\langle \Sigma^* \rangle\!\rangle$ may only be applied to proper power series. The language-theoretic counterpart of proper power series are languages in Σ^+, i.e. languages that do not contain the empty word. Therefore, one needs to justify that the rational language L can be constructed in such a way that the Kleene star is only applied to languages in Σ^+. This can be shown by structural induction (cf. Berstel and Reutenauer [2, Chapter III, Proof of Lemma 1.4] and Droste [11, Fact 6.5]). Having ensured this, the correspondences (1), (3) and (4) from Section 4.1 imply that the characteristic series $\mathbb{1}_L$ belongs to the set $\mathbb{B}^{\mathrm{rat}}\langle\!\langle \Sigma^* \rangle\!\rangle$ of rational power series. $\quad\square$

Thus, the rational power series with coefficients in the Boolean semiring $\mathbb{B}$ correspond precisely to the rational languages. If S is an arbitrary semiring with $1 + 1 \neq 1$, then e.g. the correspondence (1) from Section 4.1 does not hold true. Indeed, given two languages L_1, L_2 over Σ such that $L_1 \cap L_2 \neq \emptyset$, we obtain $\mathbb{1}_{L_1 \cup L_2} \neq \mathbb{1}_{L_1} + \mathbb{1}_{L_2}$, since

$$(\mathbb{1}_{L_1 \cup L_2}, w) = 1 \neq 1 + 1 = (\mathbb{1}_{L_1}, w) + (\mathbb{1}_{L_2}, w)$$

for any $w \in L_1 \cap L_2$. However, we have

$$\mathbb{1}_{L_1 \cup L_2} = \mathbb{1}_{L_1 \setminus L_2} + \mathbb{1}_{L_2 \setminus L_1}.$$

If we further suppose that L_1 and L_2 are rational languages, then the languages $L_1 \setminus L_2$ and $L_2 \setminus L_1$ are again rational by Corollary 2.3.8. Thus, the characteristic series $\mathbb{1}_{L_1 \cup L_2}$ of the rational language $L_1 \cup L_2$ belongs to $S^{\mathrm{rat}}\langle\!\langle \Sigma^* \rangle\!\rangle$. More generally, it can be shown that the notion of rational power series extends the language-theoretic notion of rationality in the following sense (cf. [2, Chapter III, Proposition 2.1]):

4.2.4 Proposition. *The injective map*

$$\mathrm{char}\colon \mathcal{P}(\Sigma^*) \to S\langle\!\langle \Sigma^* \rangle\!\rangle, \quad L \mapsto \mathbb{1}_L$$

preserves rationality for any semiring S, i.e. if L is a rational language over Σ, then its characteristic series $\mathbb{1}_L$ belongs to $S^{\mathrm{rat}}\langle\!\langle \Sigma^* \rangle\!\rangle$ for any semiring S.

This result implies that every rational language can be understood as a rational (unambiguous) power series in $S^{\mathrm{rat}}\langle\!\langle \Sigma^* \rangle\!\rangle$ for any semiring S (by identifying it with its characteristic series). Hence, the rational power series generalize the rational languages just as the recognizable power series generalize the recognizable languages (see Proposition 3.2.7 and Proposition 3.2.8).

4.2.5 Remark. Proposition 4.2.4 implies that the characteristic series of any rational language is again rational. One might ask whether the converse is true as well, i.e. whether the support of any rational power series is again rational. In general, this question has to be answered in the negative (cf. [2, Chapter III, Example 3.1]). However, in [2, Chapter III], Droste and Kuske [12, §9], Droste [11, §7], Salomaa and Soittola [35, §II.2] one finds various results considering the (rationality of) supports of series in $S^{\mathrm{rat}}\langle\!\langle \Sigma^* \rangle\!\rangle$. Among these, one finds the following central result, which was first proved by Schützenberger [37] for fields and later by Sontag [40] for rings (cf. [2, page 51]): *Let S be a commutative ring and let $r \in S^{\mathrm{rat}}\langle\!\langle \Sigma^* \rangle\!\rangle$ be a rational formal power series with finite image, i.e. having only finitely many distinct coefficients. Then for any subset $A \subseteq S$ the preimage*

$$r^{-1}(A) = \{w \in \Sigma^* \mid (r, w) \in A\}$$

is a rational language.

In particular, this result implies that the support of any rational power series r is a rational language, since it is given by $\mathrm{supp}(r) = r^{-1}(S \setminus \{0\})$.

4.2.6 Example. Following [2, Chapter I, Example 4.1], we consider the polynomial

$$\mathbb{1}_\Sigma = \sum_{a \in \Sigma} 1a \in S\langle \Sigma^* \rangle \subseteq S^{\mathrm{rat}}\langle\!\langle \Sigma^* \rangle\!\rangle.$$

The series $\mathbb{1}_\Sigma$ is clearly proper, and we obtain

$$(\mathbb{1}_\Sigma)^* = \sum_{n \in \mathbb{N}} (\mathbb{1}_\Sigma)^n = \sum_{w \in \Sigma^*} 1w = \mathbb{1}_{\Sigma^*},$$

since for any $n \in \mathbb{N}$ we have

$$(\mathbb{1}_\Sigma)^n = \sum_{w \in \Sigma^n} 1w = \mathbb{1}_{\Sigma^n}.$$

Thus, the series $\mathbb{1}_{\Sigma^*} = (\mathbb{1}_\Sigma)^*$ is rational. To prove $\mathbb{1}_{\Sigma^*} \in S^{\mathrm{rat}}\langle\!\langle \Sigma^* \rangle\!\rangle$, we could alternatively have applied Proposition 4.2.4 to the rational language Σ^*.

Now we consider a fixed letter $a \in \Sigma$ and the power series

$$\mathbb{1}_{\Sigma^*} \cdot 1a \cdot \mathbb{1}_{\Sigma^*}.$$

As a Cauchy product of rational power series, this is again a rational power series. Moreover, we compute

$$(\mathbb{1}_{\Sigma^*} \cdot 1a \cdot \mathbb{1}_{\Sigma^*}, w) = \sum_{\substack{u,x,v \in \Sigma^* \\ w=uvx}} \underbrace{(\mathbb{1}_{\Sigma^*}, u)}_{=1} \cdot \underbrace{(1a, x)}_{\substack{=0 \\ \text{if } x \neq a}} \cdot \underbrace{(\mathbb{1}_{\Sigma^*}, v)}_{=1} = \sum_{\substack{u,v \in \Sigma^* \\ w=uav}} 1$$

for any $w \in \Sigma^*$. Thus, the power series $\mathbb{1}_{\Sigma^*} \cdot 1a \cdot \mathbb{1}_{\Sigma^*}$ counts how often the letter a occurs in a word, where the term "how often" depends on the underlying semiring S. If we consider e.g. the semiring $\mathbb{N}$ of natural numbers, then we obtain

$$(\mathbb{1}_{\Sigma^*} \cdot 1a \cdot \mathbb{1}_{\Sigma^*}, w) = \sum_{\substack{u,v \in \Sigma^* \\ w=uav}} 1 = |w|_a$$

for any $w \in \Sigma^*$, where $|w|_a$ denotes the number of occurrences of the letter a in the word w. Indeed, each occurrence of the letter a in w corresponds to a factorization of the form $w = uav$ with $u, v \in \Sigma^*$. This yields

$$\mathbb{1}_{\Sigma^*} \cdot 1a \cdot \mathbb{1}_{\Sigma^*} = \sum_{w \in \Sigma^*} |w|_a w$$

i.e. the series $|\cdot|_a \colon \Sigma^* \to \mathbb{N}$, $w \mapsto |w|_a$ is rational. We have already encountered this power series in Example 3.2.11, where we showed that it is recognizable (over the alphabet $\Sigma = \{a, b\}$). The support of this series is given by the language

$$\{w \in \Sigma^* \mid |w|_a > 0\} = \Sigma^*\{a\}\Sigma^*,$$

which is rational and contains the infinite language $\{a\}^+ = \{a^n \mid n \in \mathbb{N}^+\}$. This shows in particular that the support of a rational power series may be infinite.

4.2.7 Remark. Assume that S is a field and consider the alphabet $\Sigma = \{x\}$. We have already indicated in Remark 4.1.5 that $S\langle\langle \{x\}^* \rangle\rangle$ and $S\langle \{x\}^* \rangle$ coincide with the rings $S[[x]]$ and $S[x]$, respectively. In [2, Chapter IV, Proposition 1.1], Berstel and Reutenauer establish a correspondence between rational power series and rational functions in one variable, as studied in Algebra. More precisely, they prove: *A formal power series $r \in S[[x]]$ is rational if and only if there exist polynomials $P, Q \in S[x]$ with $Q(0) = 1$, where $Q(0)$ denotes the coefficient (Q, ε), such that r is the power series expansion of the rational function $\frac{P}{Q}$, where $\frac{P}{Q} = P \cdot (1 - Q')^{-1}$ (cf. Neumann sum) for $Q' = 1\varepsilon - Q$.*

This result justifies the choice of Eilenberg [14] to call *rational* what – at least in Language Theory – was commonly referred to as *regular* in the foregoing literature. However, in Language Theory the terminology is still indisputable (cf. Sakarovitch [34, page 119 and page 169]).

A detailed survey on the properties of rational power series over a one-letter alphabet, also called *rational sequences*, is given in [2, Chapter IV], [14, Chapter VIII] as well as in Salomaa and Soittola [35, § II.9 and § II.10].

4.3. The Kleene–Schützenberger Theorem

Reformulating Kleene's Theorem from Section 2.3 in terms of formal power series with coefficients in the Boolean semiring $\mathbb{B}$ yields the equality

$$\mathbb{B}^{\mathrm{rat}}\langle\langle \Sigma^* \rangle\rangle = \mathbb{B}^{\mathrm{rec}}\langle\langle \Sigma^* \rangle\rangle,$$

as the power series in $\mathbb{B}^{\mathrm{rat}}\langle\langle \Sigma^* \rangle\rangle$ and $\mathbb{B}^{\mathrm{rec}}\langle\langle \Sigma^* \rangle\rangle$ correspond precisely to the rational and recognizable languages, respectively (see Proposition 4.2.3 and Proposition 3.2.7). The aim of this section is to prove this equality for arbitrary semirings.

4.3.1 Theorem (Schützenberger). *Let $r \in S\langle\langle \Sigma^* \rangle\rangle$ be a formal power series. Then r is rational if and only if r is recognizable. This means $S^{\mathrm{rat}}\langle\langle \Sigma^* \rangle\rangle = S^{\mathrm{rec}}\langle\langle \Sigma^* \rangle\rangle$.*

This result was first proved by Schützenberger [37] in 1961 for the ring $\mathbb{Z}$ of integers. The year after, he extended his result to arbitrary rings and, more generally, to semirings in [38]. As Theorem 4.3.1 contains Kleene's classical result on the equivalence of the notions of recognizability and rationality in the context of languages as a special case, it is usually called Kleene–Schützenberger Theorem, and so do we. Some authors also refer to Theorem 4.3.1 as Representation Theorem of Schützenberger (cf. Salomaa and Soittola [35, Chapter II]) or as Fundamental Theorem (cf. Sakarovitch [34, § 3.2]). Indeed, Theorem 4.3.1 is fundamental in almost all considerations concerning rational and recognizable series, respectively (cf. [35, page 20]).

Closure Properties of Recognizable Formal Power Series

We start by proving the inclusion

$$S^{\mathrm{rat}}\langle\langle \Sigma^* \rangle\rangle \subseteq S^{\mathrm{rec}}\langle\langle \Sigma^* \rangle\rangle.$$

Following Droste and Kuske [12, § 4.1], we proceed by structural induction to prove that any rational power series is recognizable. To this end, we show that the rational operations on formal power series preserve recognizability. Moreover, we prove that, under certain commutativity assumptions, $S^{\text{rec}}\langle\langle\Sigma^*\rangle\rangle$ is also closed under the Hadamard product. Thus, we extend the closure properties of recognizable languages from Proposition 2.3.4 to the realm of formal power series.

In Chapter 3 we have introduced two different modeling approaches for recognizable formal power series: weighted automata and linear representations. Therefore, to prove the recognizability of a power series, we have two possibilities: Either we choose the purely automata-theoretic approach and construct a suitable weighted automaton, or we choose the more algebraic formalism and handle linear representations. To show that any monomial is contained in $S^{\text{rec}}\langle\langle\Sigma^*\rangle\rangle$ we choose to construct suitable weighted automata.

4.3.2 Proposition. *Every monomial in $S\langle\langle\Sigma^*\rangle\rangle$ is recognizable.*

Proof. First, we consider monomials of the form $s\varepsilon$ with $s \in S$. Then the weighted automaton $\mathcal{A}_{s\varepsilon}$ which is represented by the state diagram

$$s \longrightarrow \boxed{q} \longrightarrow 1$$

fulfills $\|\mathcal{A}_{s\varepsilon}\| = s\varepsilon$. Now we consider monomials of the form sw, where $s \in S$ and $w = a_1 \dots a_n \in \Sigma^+$ is a non-empty word. Then the weighted automaton $\mathcal{A}_{sw}$ which is represented by the state diagram

$$s \longrightarrow \boxed{q_0} \xrightarrow[1]{a_1} \boxed{q_1} \xrightarrow[1]{a_2} \cdots \xrightarrow[1]{a_n} \boxed{q_n} \longrightarrow 1$$

fulfills $\|\mathcal{A}_{sw}\| = sw$. Hence, we obtain $m \in S^{\text{rec}}\langle\langle\Sigma^*\rangle\rangle$ for any monomial m in $S\langle\langle\Sigma^*\rangle\rangle$. □

To prove that $S^{\text{rec}}\langle\langle\Sigma^*\rangle\rangle$ is closed under the rational operations sum, Cauchy product and Kleene star, we employ the algebraic approach and handle linear representations. Thus, we will be confronted with matrices, in particular *blocks* of matrices. Given matrices M and N with entries coming from a semiring, we can compute the blocks of their sum $M + N$ and their product $M \cdot N$ from the blocks of M and N in the usual way (cf. Droste and Kuich [9, page 19]).

4.3.3 Proposition. *The sum of two recognizable power series is again recognizable.*

Proof. Let $r_1, r_2 \in S^{\text{rec}}\langle\langle\Sigma^*\rangle\rangle$ be recognizable power series. By Proposition 3.3.8 there exists a linear representation $(\lambda^i, \mu^i, \gamma^i)$ of dimension Q_i such that $r_i = \|(\lambda^i, \mu^i, \gamma^i)\|$ for $i = 1, 2$. Without loss of generality, we assume that the sets Q_1 and Q_2 are disjoint (otherwise, replace Q_i by the set $Q_i \times \{i\}$ for $i = 1, 2$). Then we denote by Q their disjoint union, i.e. $Q = Q_1 \cup Q_2$. Our aim is to specify a linear representation (λ, μ, γ) of dimension Q such that $r_1 + r_2 = \|(\lambda, \mu, \gamma)\|$.

To this end, we define a row vector $\lambda \in S^Q$ and a column vector $\gamma \in S^Q$ by

$$\lambda = \begin{pmatrix} \lambda^1 & \lambda^2 \end{pmatrix} \quad \text{and} \quad \gamma = \begin{pmatrix} \gamma^1 \\ \gamma^2 \end{pmatrix}.$$

Furthermore, for each $w \in \Sigma^*$ we define a matrix $\mu(w) \in S^{Q \times Q}$ by

$$\mu(w) = \begin{pmatrix} \mu^1(w) & 0 \\ 0 & \mu^2(w) \end{pmatrix}.$$

It is straightforward to check that the map $\mu \colon \Sigma^* \to S^{Q \times Q}$, $w \mapsto \mu(w)$ constitutes a monoid homomorphism. Hence, (λ, μ, γ) is a linear representation of dimension Q. Furthermore, we obtain

$$\lambda \cdot \mu(w) \cdot \gamma = \lambda^1 \cdot \mu^1(w) \cdot \gamma^1 + \lambda^2 \cdot \mu^2(w) \cdot \gamma^2 = (r_1, w) + (r_2, w)$$

for any $w \in \Sigma^*$. Therefore, $r_1 + r_2 = ||(\lambda, \mu, \gamma)||$ is recognizable by Proposition 3.3.8. $\square$

4.3.4 Remark. We want to sketch an alternative proof of Proposition 4.3.3, employing the purely automata-theoretic approach. Thus, let $\mathcal{A}_i = (Q_i, \mathrm{in}_i, \mathrm{wt}_i, \mathrm{out}_i)$ be a weighted automaton for $i = 1, 2$. Without loss of generality, assume again that the state sets Q_1 and Q_2 are disjoint (otherwise, replace Q_i by the set $Q_i \times \{i\}$ for $i = 1, 2$). Then we define a weighted automaton $\mathcal{A}_1 \oplus \mathcal{A}_2 = (Q, \mathrm{in}, \mathrm{wt}, \mathrm{out})$, having as state set the disjoint union $Q = Q_1 \cup Q_2$, by the functions

$$\mathrm{in}(q) = \begin{cases} \mathrm{in}_1(q) & \text{if } q \in Q_1 \\ \mathrm{in}_2(q) & \text{if } q \in Q_2 \end{cases} \quad (q \in Q),$$

$$\mathrm{wt}(p, a, q) = \begin{cases} \mathrm{wt}_1(p, a, q) & \text{if } p, q \in Q_1 \\ \mathrm{wt}_2(p, a, q) & \text{if } p, q \in Q_2 \quad (p, q \in Q, a \in \Sigma), \\ 0 & \text{otherwise} \end{cases}$$

$$\mathrm{out}(q) = \begin{cases} \mathrm{out}_1(q) & \text{if } q \in Q_1 \\ \mathrm{out}_2(q) & \text{if } q \in Q_2 \end{cases} \quad (q \in Q).$$

The automaton $\mathcal{A}_1 \oplus \mathcal{A}_2$ is called **disjoint union** of $\mathcal{A}_1$ and $\mathcal{A}_2$, and it can be shown that its behavior is given by $||\mathcal{A}_1 \oplus \mathcal{A}_2|| = ||\mathcal{A}_1|| + ||\mathcal{A}_2||$ (cf. Droste [11, Lemma 6.10]).

As every polynomial is a finite sum of monomials, we infer from Proposition 4.3.2 and Proposition 4.3.3 that every polynomial is recognizable, i.e. $S\langle \Sigma^* \rangle \subseteq S^{\mathrm{rec}}\langle\langle \Sigma^* \rangle\rangle$.

4.3.5 Proposition. *The Cauchy product of two recognizable power series is again recognizable.*

Proof. Let $r_1, r_2 \in S^{\mathrm{rec}}\langle\langle \Sigma^* \rangle\rangle$ be recognizable power series. By Proposition 3.3.8 there exists a linear representation $(\lambda^i, \mu^i, \gamma^i)$ of dimension Q_i such that $r_i = ||(\lambda^i, \mu^i, \gamma^i)||$ for $i = 1, 2$. Without loss of generality, we assume that the sets Q_1 and Q_2 are disjoint (otherwise, replace Q_i by the set $Q_i \times \{i\}$ for $i = 1, 2$). Then we denote by Q their disjoint union, i.e. $Q = Q_1 \cup Q_2$. Our aim is to specify a linear representation (λ, μ, γ) of dimension Q such that $r_1 \cdot r_2 = ||(\lambda, \mu, \gamma)||$.

To this end, we define a row vector $\lambda \in S^Q$ and a column vector $\gamma \in S^Q$ by

$$\lambda = \begin{pmatrix} \lambda^1 & 0 \end{pmatrix} \quad \text{and} \quad \gamma = \begin{pmatrix} \gamma^1 \lambda^2 \gamma^2 \\ \gamma^2 \end{pmatrix}.$$

Furthermore, for each $w \in \Sigma^*$ we define a matrix $\mu(w) \in S^{Q \times Q}$ by

$$\mu(w) = \begin{pmatrix} \mu^1(w) & \sum\limits_{\substack{w=uv \\ v \neq \varepsilon}} \mu^1(u)\gamma^1\lambda^2\mu^2(v) \\ 0 & \mu^2(w) \end{pmatrix}.$$

Claim. *The map* $\mu \colon \Sigma^* \to S^{Q \times Q}$, $w \mapsto \mu(w)$ *constitutes a monoid homomorphism.*

Proof of Claim. First, we consider the empty word and show $\mu(\varepsilon) = E_Q$. Since μ^i is a monoid homomorphism, we have $\mu^i(\varepsilon) = E_{Q_i}$ for $i = 1, 2$. Thus, we obtain

$$\mu(\varepsilon) = \begin{pmatrix} \mu^1(\varepsilon) & \sum\limits_{\substack{\varepsilon=uv \\ v \neq \varepsilon}} \mu^1(u)\gamma^1\lambda^2\mu^2(v) \\ 0 & \mu^2(\varepsilon) \end{pmatrix} = \begin{pmatrix} E_{Q_1} & 0 \\ 0 & E_{Q_2} \end{pmatrix} = E_Q,$$

as empty sums evaluate to 0 (see Notation 3.1.3). Now let $w_1, w_2 \in \Sigma^*$. Our aim is to show that $\mu(w_1 w_2) = \mu(w_1) \cdot \mu(w_2)$. Since μ^i is a monoid homomorphism for $i = 1, 2$, we have $\mu^i(w_1 w_2) = \mu^i(w_1)\mu^i(w_2)$. Thus, by definition of μ we obtain

$$\mu(w_1 w_2) = \begin{pmatrix} \mu^1(w_1)\mu^1(w_2) & \sum\limits_{\substack{w_1 w_2=uv \\ v \neq \varepsilon}} \mu^1(u)\gamma^1\lambda^2\mu^2(v) \\ 0 & \mu^2(w_1)\mu^2(w_2) \end{pmatrix}.$$

On the other hand, we compute

$$\mu(w_1) \cdot \mu(w_2) = \begin{pmatrix} \mu^1(w_1)\mu^1(w_2) & M \\ 0 & \mu^2(w_1)\mu^2(w_2) \end{pmatrix},$$

where the matrix $M \in S^{Q_1 \times Q_2}$ is given by

$$M = \mu^1(w_1) \cdot \left(\sum_{\substack{w_2=u_2 v_2 \\ v_2 \neq \varepsilon}} \mu^1(u_2)\gamma^1\lambda^2\mu^2(v_2) \right)$$

$$+ \left(\sum_{\substack{w_1=u_1 v_1 \\ v_1 \neq \varepsilon}} \mu^1(u_1)\gamma^1\lambda^2\mu^2(v_1) \right) \cdot \mu^2(w_2).$$

Hence, to establish the equality $\mu(w_1 w_2) = \mu(w_1) \cdot \mu(w_2)$, it remains to justify that

$$M = \sum_{\substack{w_1 w_2=uv \\ v \neq \varepsilon}} \mu^1(u)\gamma^1\lambda^2\mu^2(v).$$

We first observe that any factorization $w_1 w_2 = uv$ with $u, v \in \Sigma^*$ fulfills exactly one of the following two conditions:

(i) w_1 is a prefix of u, i.e. $u = w_1 u'$ for some $u' \in \Sigma^*$.

(ii) w_2 is a proper suffix of v, i.e. $v = v' w_2$ for some $v' \in \Sigma^*$ with $v' \neq \varepsilon$.

Condition (ii) implies in particular that $v \neq \varepsilon$, as v contains the non-empty word v'. This yields

$$
\begin{aligned}
M &= \sum_{\substack{w_2 = u_2 v_2 \\ v_2 \neq \varepsilon}} \mu^1(w_1)\mu^1(u_2)\gamma^1\lambda^2\mu^2(v_2) + \sum_{\substack{w_1 = u_1 v_1 \\ v_1 \neq \varepsilon}} \mu^1(u_1)\gamma^1\lambda^2\mu^2(v_1)\mu^2(w_2) \\
&= \sum_{\substack{w_2 = u_2 v_2 \\ v_2 \neq \varepsilon}} \underbrace{\mu^1(w_1 u_2)\gamma^1\lambda^2\mu^2(v_2)}_{\substack{w_1 w_2 = uv \\ \text{with } u = w_1 u_2, v = v_2 \neq \varepsilon \\ \text{fulfills condition } (i)}} + \sum_{\substack{w_1 = u_1 v_1 \\ v_1 \neq \varepsilon}} \underbrace{\mu^1(u_1)\gamma^1\lambda^2\mu^2(v_1 w_2)}_{\substack{w_1 w_2 = uv \\ \text{with } u = u_1, v = v_1 w_2 \neq \varepsilon \\ \text{fulfills condition } (ii)}} \\
&= \sum_{\substack{w_1 w_2 = uv \\ v \neq \varepsilon}} \mu^1(u)\gamma^1\lambda^2\mu^2(v).
\end{aligned}
$$

Therefore, μ is compatible with the multiplications of the monoids Σ^* and $S^{Q \times Q}$, completing the proof of the claim. ◇

Hence, (λ, μ, γ) is a linear representation of dimension Q. Furthermore, we obtain

$$
\begin{aligned}
\lambda \cdot \mu(w) \cdot \gamma &= \lambda^1 \mu^1(w)\gamma^1\lambda^2\gamma^2 + \lambda^1 \left(\sum_{\substack{w = uv \\ v \neq \varepsilon}} \mu^1(u)\gamma^1\lambda^2\mu^2(v) \right) \gamma^2 \\
&= (r_1, w)(r_2, \varepsilon) + \sum_{\substack{w = uv \\ v \neq \varepsilon}} \lambda^1 \mu^1(u)\gamma^1\lambda^2\mu^2(v)\gamma^2 \\
&= (r_1, w)(r_2, \varepsilon) + \sum_{\substack{w = uv \\ v \neq \varepsilon}} (r_1, u)(r_2, v) \\
&= \sum_{w = uv} (r_1, u)(r_2, v) \\
&= (r_1 \cdot r_2, w)
\end{aligned}
$$

for any $w \in \Sigma^*$. Therefore, the Cauchy product $r_1 \cdot r_2$ coincides with the formal power series $||(\lambda, \mu, \gamma)||$, which is recognizable by Proposition 3.3.8. □

4.3.6 Proposition. *The Kleene star of a proper recognizable power series is again recognizable.*

Proof. Let $r \in S^{\mathrm{rec}}\langle\!\langle \Sigma^* \rangle\!\rangle$ be a proper recognizable power series. By Proposition 3.3.8 there exists a linear representation (λ, μ, γ) of dimension Q such that $r = ||(\lambda, \mu, \gamma)||$. We consider the map

$$\Sigma \to S^{Q \times Q}, \quad a \mapsto \mu(a) + \gamma\lambda\mu(a),$$

and we denote by μ' the unique monoid homomorphism μ' from the free monoid Σ^* into the multiplicative monoid $(S^{Q \times Q}, \cdot, E_Q)$ of matrices, which extends this map (see Lemma 2.1.12).

Claim. *We have*

$$\mu'(w) = \sum_{\substack{w=w_1\ldots w_k \\ k\in\mathbb{N}^+,\, w_i\in\Sigma^+}} \left((\mu(w_1) + \gamma\lambda\mu(w_1)) \cdot \prod_{i=2}^{k} \gamma\lambda\mu(w_i) \right)$$

for any non-empty word $w = a_1 \ldots a_n \in \Sigma^+$.

Proof of Claim. We proceed by induction on $|w| = n$. Recall that $\mu(w)$ is given by the product

$$\mu'(w) = \mu'(a_1 \ldots a_n) = \prod_{j=1}^{n} \mu'(a_j) = \prod_{j=1}^{n} (\mu(a_j) + \gamma\lambda\mu(a_j)).$$

Thus, it is straightforward to verify the claim for $n = 1$, as empty products evaluate to 1 (see Notation 3.1.3). We consider now a non-empty word $w = a_1 \ldots a_n \in \Sigma^+$ of length $|w| = n \geq 2$, and we assume that the claim holds for all words of length $n - 1$. Then we set $w' = a_1 \ldots a_{n-1}$, and we observe that any factorization $w = w_1 \ldots w_k$ with $k \in \mathbb{N}^+$ and non-empty words $w_1, \ldots, w_k \in \Sigma^+$ fulfills exactly one of the following conditions:

(i) The last letter a_n of w is a proper suffix of the last subword w_k, i.e. we have $w_k = w_k' a_n$ for some non-empty word $w_k' \neq \varepsilon$.

(ii) The last letter a_n of w coincides with the last subword w_k, i.e. we have $w_k = a_n$.

Moreover, we infer from condition (i) that the subword w' is given by $w' = w_1 \ldots w_{k-1}$, and condition (ii) implies that $w' = w_1 \ldots w_{k-1} w_k'$. Hence, applying the induction hypothesis yields

$$\mu'(w) = \mu'(w') \cdot (\mu(a_n) + \gamma\lambda\mu(a_n))$$

$$= \sum_{\substack{w'=w_1'\ldots w_k' \\ k\in\mathbb{N}^+,\, w_i'\in\Sigma^+}} \left((\mu(w_1') + \gamma\lambda\mu(w_1')) \cdot \left(\prod_{i=2}^{k} \gamma\lambda\mu(w_i') \right) \cdot \mu(a_n) \right)$$

$$+ \sum_{\substack{w'=w_1'\ldots w_k' \\ k\in\mathbb{N}^+,\, w_i'\in\Sigma^+}} \left((\mu(w_1') + \gamma\lambda\mu(w_1')) \cdot \left(\prod_{i=2}^{k} \gamma\lambda\mu(w_i') \right) \cdot \gamma\lambda\mu(a_n) \right)$$

$$= \sum_{\substack{w'=w_1'\ldots w_k' \\ k\in\mathbb{N}^+,\, w_i'\in\Sigma^+}} \left((\mu(w_1') + \gamma\lambda\mu(w_1')) \cdot \left(\prod_{i=2}^{k-1} \gamma\lambda\mu(w_i') \right) \cdot \gamma\lambda \underbrace{\mu(w_k')\mu(a_n)}_{=\mu(w_k'a_n)} \right)$$

$$+ \sum_{\substack{w'=w_1'\ldots w_k' \\ k\in\mathbb{N}^+,\, w_i'\in\Sigma^+}} \left((\mu(w_1') + \gamma\lambda\mu(w_1')) \cdot \left(\prod_{i=2}^{k} \gamma\lambda\mu(w_i') \right) \cdot \gamma\lambda\mu(a_n) \right)$$

$$= \sum_{\substack{w=w_1\ldots w_k \\ k\in\mathbb{N}^+,\, w_i\in\Sigma^+}} \left((\mu(w_1) + \gamma\lambda\mu(w_1)) \cdot \prod_{i=2}^{k} \gamma\lambda\mu(w_i) \right).$$

◇

Considering the empty word, we have $\mu(\varepsilon) = E_Q$ and thus

$$\lambda\gamma = \lambda E_Q \gamma = \lambda\mu(\varepsilon)\gamma = (r,\varepsilon) = 0,$$

since the power series r is proper. In particular, this yields

$$\lambda\mu'(\varepsilon)\gamma = \lambda\gamma = 0.$$

On the other hand, if we exploit the identity for non-empty words, which we have established above, then we obtain

$$\lambda\mu'(w)\gamma = \sum_{\substack{w=w_1\ldots w_k \\ k\in\mathbb{N}^+, w_i\in\Sigma^+}} \lambda \cdot \left((\mu(w_1) + \gamma\lambda\mu(w_1)) \cdot \prod_{i=2}^{k} \gamma\lambda\mu(w_i)\right) \cdot \gamma$$

$$= \sum_{\substack{w=w_1\ldots w_k \\ k\in\mathbb{N}^+, w_i\in\Sigma^+}} (\lambda\mu(w_1) + \underbrace{\lambda\gamma\lambda\mu(w_1)}_{\substack{=0 \\ \text{since } \lambda\gamma=0}}) \cdot \left(\prod_{i=2}^{k} \gamma\lambda\mu(w_i)\right) \cdot \gamma$$

$$= \sum_{\substack{w=w_1\ldots w_k \\ k\in\mathbb{N}^+, w_i\in\Sigma^+}} \lambda\mu(w_1)\gamma \cdot \prod_{i=2}^{k} \lambda\mu(w_i)\gamma$$

$$= \sum_{\substack{w=w_1\ldots w_k \\ k\in\mathbb{N}^+, w_i\in\Sigma^+}} \prod_{i=1}^{k} \underbrace{\lambda\mu(w_i)\gamma}_{=(r,w_i)}$$

$$= (r^*, w)$$

for any non-empty word $w = a_1\ldots a_n \in \Sigma^+$. Therefore, the star r^* coincides with the power series $\|(\lambda,\mu',\gamma)\| + 1\varepsilon$, which is recognizable by Proposition 3.3.8, Proposition 4.3.2 and Proposition 4.3.3. □

By Proposition 4.3.3, Proposition 4.3.5 and Proposition 4.3.6, the set $S^{\text{rec}}\langle\!\langle\Sigma^*\rangle\!\rangle$ of recognizable power series is closed under the rational operations sum, Cauchy product and Kleene star. Under certain commutativity assumptions, even the Hadamard product preserves recognizability. In order to state the corresponding result, we first introduce further notions in the context of semirings based on Droste and Gastin [10, page 179].

4.3.7 Definition. Let S be a semiring.

a) A **subsemiring** of S is a subset $S' \subseteq S$ containing 0 and 1 which is closed under addition and multiplication, that is, $s_1 + s_2 \in S'$ and $s_1 \cdot s_2 \in S'$ for any $s_1, s_2 \in S'$.

b) Let $M \subseteq S$ be a subset of S. We denote by $\langle M\rangle$ the **subsemiring generated by** M, which is given by

$$\langle M\rangle := \left\{\sum_{i=1}^{n} s_i \,\middle|\, n \in \mathbb{N}, s_1,\ldots,s_n \in \langle M\rangle_{\text{mult}}\right\},$$

where $\langle M \rangle_{\text{mult}}$ is defined by

$$\langle M \rangle_{\text{mult}} := \left\{ \prod_{i=1}^{n} s_i \;\middle|\; n \in \mathbb{N}, s_1, \ldots, s_n \in M \right\}.$$

It is straightforward to verify that $\langle M \rangle$ indeed constitutes a subsemiring of S. One can even show that it is the smallest subsemiring containing the set M.

4.3.8 Definition. Let S be a semiring and $M_1, M_2 \subseteq S$ be subsets of S. We say that M_1 and M_2 **commute elementwise** if $s_1 \cdot s_2 = s_2 \cdot s_1$ for any $s_1 \in M_1$ and $s_2 \in M_2$.

The distributivity in S implies that the subsets M_1 and M_2 commute elementwise if and only if the sets $\langle M_1 \rangle$ and $\langle M_2 \rangle$ commute elementwise (cf. [10, page 179]).

4.3.9 Example. The set $S^{\text{rat}}\langle\langle \Sigma^* \rangle\rangle$ of rational power series is a subsemiring of the semiring $S\langle\langle \Sigma^* \rangle\rangle$. More precisely, it is the smallest subsemiring of $S\langle\langle \Sigma^* \rangle\rangle$ that contains the set $S\langle \Sigma^* \rangle$ of polynomials and is closed under the Kleene star (applied to proper series only).

4.3.10 Proposition. *Let S_1, S_2 be subsemirings of S such that S_1 and S_2 commute elementwise. If $r_1 \in S_1^{\text{rec}}\langle\langle \Sigma^* \rangle\rangle$ and $r_2 \in S_2^{\text{rec}}\langle\langle \Sigma^* \rangle\rangle$, then $r_1 \odot r_2 \in S^{\text{rec}}\langle\langle \Sigma^* \rangle\rangle$.*

Proof. We sketch the proof following [12, Lemma 4.3]. Let $\mathcal{A}_i = (Q_i, \text{in}_i, \text{wt}_i, \text{out}_i)$ be a weighted automaton with weights in S_i and behavior $\|\mathcal{A}_i\| = r_i$ for $i = 1, 2$. We define a weighted automaton $\mathcal{A} = (Q, \text{in}, \text{wt}, \text{out})$ with state set $Q = Q_1 \times Q_2$ and weights in S by

$$\text{in}(q) = \text{in}(q_1) \cdot \text{in}(q_2),$$
$$\text{wt}(p, a, q) = \text{wt}_1(p_1, a, q_1) \cdot \text{wt}_2(p_2, a, q_2),$$
$$\text{out}(q) = \text{out}(q_1) \cdot \text{out}(q_2),$$

where $p = (p_1, p_2), q = (q_1, q_2) \in Q$ and $a \in \Sigma$. Then, exploiting that S_1 and S_2 commute elementwise, it can be verified that

$$\|\mathcal{A}\| = \|\mathcal{A}_1\| \odot \|\mathcal{A}_2\| = r_1 \odot r_2.$$

Considering for instance the empty word and using distributivity, we obtain

$$(\|\mathcal{A}\|, \varepsilon) = \sum_{q=(q_1,q_2)\in Q} \text{in}(q) \cdot \text{out}(q)$$
$$= \sum_{q=(q_1,q_2)\in Q} \text{in}(q_1) \cdot \text{in}(q_2) \cdot \text{out}(q_1) \cdot \text{out}(q_2)$$
$$= \sum_{q=(q_1,q_2)\in Q} \text{in}(q_1) \cdot \text{out}(q_1) \cdot \text{in}(q_2) \cdot \text{out}(q_2)$$
$$= \sum_{q_1 \in Q_1} \text{in}(q_1)\,\text{out}(q_1) \cdot \sum_{q_2 \in Q_2} \text{in}(q_2)\,\text{out}(q_2)$$

$$= (||\mathcal{A}_1||, \varepsilon) \cdot (||\mathcal{A}_2||, \varepsilon)$$
$$= (||\mathcal{A}_1|| \odot ||\mathcal{A}_2||, \varepsilon),$$

since S_1 and S_2 commute elementwise. In [12, Lemma 4.3] one finds a similar justification for letters in Σ. A more detailed proof of the proposition can be found in Droste [11, Theorem 6.20]. □

4.3.11 Remark.

a) As a consequence of Proposition 4.3.10, the Hadamard product $r_1 \odot r_2$ of two recognizable power series $r_1, r_2 \in S^{\mathrm{rec}}\langle\!\langle \Sigma^* \rangle\!\rangle$ is again recognizable if the underlying semiring S is commutative. However, in general the assumption that the subsemirings S_1 and S_2 of S commute elementwise is necessary to guarantee $r_1 \odot r_2 \in S^{\mathrm{rec}}\langle\!\langle \Sigma^* \rangle\!\rangle$ (cf. [12, Example 4.2]).

b) As a consequence of Proposition 4.3.10, we obtain that certain "restrictions" of recognizable power series are still recognizable (cf. [12, Corollary 4.4]). More precisely, let $r \in S^{\mathrm{rec}}\langle\!\langle \Sigma^* \rangle\!\rangle$ be a recognizable power series, L a recognizable language over Σ and consider the semiring $S_1 = \langle\{0,1\}\rangle = \langle\emptyset\rangle$. By Proposition 3.2.8, the characteristic series $\mathbb{1}_L$ belongs to $S_1^{\mathrm{rec}}\langle\!\langle \Sigma^* \rangle\!\rangle$. Of course, the sets S and $\langle\{0,1\}\rangle$ commute elementwise, as $0 \cdot s = s \cdot 0 = 0$ and $1 \cdot s = s \cdot 1 = s$ for any $s \in S$. Thus, it follows immediately from Proposition 4.3.10 that the power series $r \odot \mathbb{1}_L$ is again recognizable. The power series $r \odot \mathbb{1}_L$ can be considered as restriction of the power series r to L, since for any $w \in \Sigma^*$ we have

$$(r \odot \mathbb{1}_L, w) = (r, w) \cdot (\mathbb{1}_L, w) = \begin{cases} (r, w) & \text{if } w \in L \\ 0 & \text{otherwise,} \end{cases}$$

which yields

$$r \odot \mathbb{1}_L = \sum_{w \in L} (r, w) w.$$

4.3.12 Remark. Using the above established closure properties of the set $S^{\mathrm{rec}}\langle\!\langle \Sigma^* \rangle\!\rangle$, we can deduce that also scalar multiplication preserves recognizability. Indeed, given a recognizable power series $r \in S^{\mathrm{rec}}\langle\!\langle \Sigma^* \rangle\!\rangle$ and $s \in S$, we have multiple opportunities to prove that the scalar product sr is again recognizable:

- By expressing the scalar product as Cauchy product $sr = s\varepsilon \cdot r$ of a monomial and the recognizable power series r, we can apply Proposition 4.3.2 and Proposition 4.3.5 to obtain $sr \in S^{\mathrm{rec}}\langle\!\langle \Sigma^* \rangle\!\rangle$.

- Alternatively, let $\mathcal{A} = (Q, \mathrm{in}, \mathrm{wt}, \mathrm{out})$ be a weighted automaton with $||\mathcal{A}|| = r$. We can replace the weight function in by the function in_s, defined by $\mathrm{in}_s(q) = s \cdot \mathrm{in}(q)$ for any $q \in Q$, to obtain a weighted automaton $\mathcal{A}_s = (Q, \mathrm{in}_s, \mathrm{wt}, \mathrm{out})$ having the scalar product sr as behavior.

- Furthermore, we can apply Proposition 3.3.8 and specify a linear representation defining sr (cf. Droste [11, Lemma 6.18]).

In particular, we infer that for any $s \in S$ the constant series

$$s \cdot 1_{\Sigma^*} = \sum_{w \in \Sigma^*} sw$$

is recognizable, as 1_{Σ^*} is recognizable by Proposition 3.2.8.

Proceeding by induction and, in doing so, using the closure properties of the set $S^{\mathrm{rec}} \langle\!\langle \Sigma^* \rangle\!\rangle$ established throughout this section, yields the inclusion

$$S^{\mathrm{rat}} \langle\!\langle \Sigma^* \rangle\!\rangle \subseteq S^{\mathrm{rec}} \langle\!\langle \Sigma^* \rangle\!\rangle.$$

Weighted Automata and Linear Systems of Equations

In this section, we aim at proving the inclusion

$$S^{\mathrm{rec}} \langle\!\langle \Sigma^* \rangle\!\rangle \subseteq S^{\mathrm{rat}} \langle\!\langle \Sigma^* \rangle\!\rangle.$$

To establish this inclusion, we exploit interconnections between weighted automata, formal power series and linear systems of equations. More precisely, following Droste and Kuske [12, § 4.2], we give an algebraic method for computing the behavior of a weighted automaton which depends on the study of linear systems of equations. First, we investigate linear equations and their solutions. We then consider linear systems of equations having rational power series as coefficients. We show in Proposition 4.3.17 that the solutions of such a system are again rational power series. Lemma 4.3.19 provides an identity for the behavior of a weighted automaton. Based on this identity, in the proof of Proposition 4.3.20 we demonstrate how a weighted automaton can be transformed into a linear system of equations with rational coefficients. This linear system will be defined in such a way that the behavior of the weighted automaton coincides with a component of the solution vector of the system. Combining these observations, we obtain a proof for the inclusion $S^{\mathrm{rec}} \langle\!\langle \Sigma^* \rangle\!\rangle \subseteq S^{\mathrm{rat}} \langle\!\langle \Sigma^* \rangle\!\rangle$.

Let $r, r' \in S \langle\!\langle \Sigma^* \rangle\!\rangle$ be formal power series. In the following, **linear equations** of the form

$$X = rX + r', \tag{1}$$

where X denotes a variable, play a central role.

4.3.13 Definition. A power series $z \in S \langle\!\langle \Sigma^* \rangle\!\rangle$ is called **solution** of the equation (1) if it satisfies $z = rz + r'$. We call a linear equation of the form (1) **proper** if the power series r is proper.

The next results shows that proper linear equations always have a unique solution.

4.3.14 Lemma. *Let $r, r' \in S \langle\!\langle \Sigma^* \rangle\!\rangle$ be formal power series. If the power series r is proper, then the linear equation $X = rX + r'$ has the unique solution $r^* r'$.*

Proof. We assume that we are given a solution $z \in S\langle\!\langle \Sigma^* \rangle\!\rangle$ of the proper linear equation $X = rX + r'$, and we consider an arbitrary $w \in \Sigma^*$. Exploiting the equation $z = rz + r'$, we can proceed inductively to obtain

$$z = rz + r'$$
$$= r(rz + r') + r' = r^2 z + rr' + r'$$
$$= r^2(rz + r') + rr' + r' = r^3 z + r^2 r' + rr' + r'$$
$$\vdots$$
$$= r^{|w|+1} z + \sum_{n=0}^{|w|} r^n r'.$$

Since we assume r to be proper, for any $u \in \Sigma^*$ and any $m > |u|$ we have $(r^m, u) = 0$. Thus, for any $u \in \Sigma^*$ we obtain

$$(r^*, u) = \sum_{n \in \mathbb{N}} (r^n, u) = \sum_{n=0}^{|u|} (r^n, u).$$

This yields

$$(z, w) = (r^{|w|+1} z, w) + \sum_{n=0}^{|w|} (r^n r', w)$$
$$= \sum_{w=uv} \underbrace{(r^{|w|+1}, u)}_{=0 \text{ since } |u| \le |w| < |w|+1} (z, v) + \sum_{n=0}^{|w|} \sum_{w=uv} (r^n, u)(r', v)$$
$$= \sum_{n=0}^{|w|} \sum_{w=uv} (r^n, u)(r', v)$$
$$= \sum_{w=uv} \left(\sum_{n=0}^{|w|} (r^n, u) \right) (r', v)$$
$$= \sum_{w=uv} (r^*, u)(r', v) = (r^* r', w).$$

Hence, the power series z coincides with the Cauchy product $r^* r'$, which indeed solves the equation $X = rX + r'$, as we have

$$rr^* r' + r' = (rr^* + 1\varepsilon)r' = (r^+ + 1\varepsilon)r' = r^* r'.$$

Consequently, $r^* r'$ is the unique solution of the proper linear equation $X = rX + r'$. $\square$

4.3.15 Definition. A **linear system (of equations)** is a system of finitely many equations of the form

$$\left(X_i = \sum_{j=0}^{n} r_{ij} X_j + r_i \right)_{0 \le i \le n} \tag{2}$$

where $n \in \mathbb{N}$, $r_{ij}, r_i \in S\langle\langle \Sigma^* \rangle\rangle$ are power series for any $i, j \in \{0, \ldots, n\}$ and X_i denotes a variable for any $i \in \{0, \ldots, n\}$. The power series r_{ij} and r_i $(i, j \in \{0, \ldots, n\})$ are called **coefficients** of the linear system (2). We call the linear system (2) **proper** if the coefficient r_{ij} is a proper power series for any $i, j \in \{0, \ldots, n\}$. Given formal power series $z_0, \ldots, z_n \in S\langle\langle \Sigma^* \rangle\rangle$, the vector $(z_1, \ldots, z_n)$ is called **solution (vector)** of the linear system (2) if it satisfies

$$z_i = \sum_{j=0}^{n} r_{ij} z_j + r_i$$

for any $i \in \{0, \ldots, n\}$.

For $n = 0$ the linear system (2) is a single linear equation of the form (1).

4.3.16 Remark. Set $Q = \{0, \ldots, n\}$. The coefficient matrix

$$R \in (S\langle\langle \Sigma^* \rangle\rangle)^{Q \times Q} \text{ with } R_{i,j} = r_{ij}$$

can be regarded as a power series with coefficients in $S^{Q \times Q}$, i.e. as an element of the set $S^{Q \times Q}\langle\langle \Sigma^* \rangle\rangle$, and one can show that this power series is proper if all entries r_{ij} $(i, j \in \{0, \ldots, n\})$ are proper power series. Hence, using matrix notation, the linear system (2) can be written as

$$X = RX + R',$$

where we consider $X = (X_0, \ldots, X_n)$ and $R' = (r_0, \ldots, r_n)$ as column vectors. Moreover, the system (2) is proper if and only if the matrix R (regarded as power series) is proper (cf. Sakarovitch [34, Remark 2.18]).

Proper linear systems are of particular interest when working with rational power series, as they preserve rationality in the following sense:

4.3.17 Proposition. *Let $n \in \mathbb{N}$ and let $r_{ij}, r_i \in S^{\mathrm{rat}}\langle\langle \Sigma^* \rangle\rangle$ be rational power series such that r_{ij} is proper for any $i, j \in \{0, \ldots, n\}$. If $(z_0, \ldots, z_n)$ is a solution of the proper linear system (2), then its components $z_0, \ldots, z_n$ are rational power series.*

Proof. We proceed by induction on n. For $n = 0$, the system (2) consists of just one proper linear equation of the form $X_0 = r_{00} X_0 + r_0$. By Lemma 4.3.14 the unique solution of this proper linear equation is given by the power series $z_0 = r_{00}^* r_0$. Therefore, the solution $z_0 = r_{00}^* r_0$ is rational, as the coefficients r_{00} and r_0 are both rational power series. We consider now a fixed $n \in \mathbb{N}^+$ and we assume that the claim holds for $n - 1$. Given a solution $(z_0, \ldots, z_n)$ of the proper linear system (2), the last component z_n then solves the proper linear equation

$$X_n = r_{nn}X_n + \sum_{j=0}^{n-1} r_{nj}z_j + r_n.$$

Thus, Lemma 4.3.14 implies

$$z_n = r_{nn}^* \left(\sum_{j=0}^{n-1} r_{nj}z_j + r_n \right).$$

Substituting this into the first n equations of (2), for any $i \in \{0, \ldots, n-1\}$ we obtain

$$z_i = \sum_{j=0}^{n-1} r_{ij}z_j + r_{in}z_n + r_i$$

$$= \sum_{j=0}^{n-1} r_{ij}z_j + r_{in}r_{nn}^* \left(\sum_{j=0}^{n-1} r_{nj}z_j + r_n \right) + r_i$$

$$= \sum_{j=0}^{n-1} \underbrace{(r_{ij} + r_{in}r_{nn}^* r_{nj})}_{=:\bar{r}_{ij}} z_j + \underbrace{r_{in}r_{nn}^* r_n + r_i}_{=:\bar{r}_i}.$$

Hence, $(z_0, \ldots, z_{n-1})$ is a solution of the linear system

$$\left(X_i = \sum_{j=0}^{n-1} \bar{r}_{ij}X_j + \bar{r}_i \right)_{0 \le i \le n-1} \tag{3}$$

whose coefficients $\bar{r}_{ij}, \bar{r}_i$ $(i, j \in \{0, \ldots, n-1\})$ are again rational power series. Moreover, for any $i, j \in \{0, \ldots, n-1\}$ the power series r_{in}, r_{nj} and r_{ij} are all proper, which implies

$$(\bar{r}_{ij}, \varepsilon) = \underbrace{(r_{ij}, \varepsilon)}_{=0} + \underbrace{(r_{in}, \varepsilon)}_{=0}(r_{nn}^*, \varepsilon)\underbrace{(r_{nj}, \varepsilon)}_{=0} = 0,$$

i.e. also the power series $\bar{r}_{ij}$ is proper. Hence, (3) constitutes a proper linear system with rational coefficients. As (3) is a linear system of only n equations and unknowns, respectively, we can apply the induction hypothesis to obtain that each component of its solution $(z_0, \ldots, z_{n-1})$ is a rational power series. Consequently, also

$$z_n = r_{nn}^* \left(\sum_{j=0}^{n-1} r_{nj}z_j + r_n \right)$$

is a rational power series. Summarizing, each component of the solution $(z_0, \ldots, z_n)$ of the original system (2) is rational, completing the induction step. $\qquad \square$

Thus, given a proper linear system with rational coefficients, the components of any solution vector are again rational. In fact, the proof of Proposition 4.3.17 shows in particular that such a system has a unique solution, as we inductively apply Lemma 4.3.14. In the following, we want to exploit these properties of linear systems, to show that every recognizable power series is rational. More precisely, we will express the behavior of a weighted automaton in terms of the solution of a proper linear system having rational power series as coefficients. It will then follow from Proposition 4.3.17 that the behavior is itself a rational power series.

4.3.18 Definition. Let $\mathcal{A} = (Q, \text{in}, \text{wt}, \text{out})$ be a weighted automaton over Σ with weights in S. For any $p \in Q$, we define a weighted automaton $\mathcal{A}_p = (Q, \text{in}_p, \text{wt}, \text{out})$, where the weight function in_p is defined by

$$\text{in}_p(q) = \begin{cases} 1 & \text{if } q = p \\ 0 & \text{otherwise} \end{cases}$$

for any $q \in Q$.

As we have indicated above, we will transform a weighted automaton $\mathcal{A}$ into a proper linear system with rational coefficients. For each state $p \in Q$, the behavior $||\mathcal{A}_p||$ will serve as component of the solution vector of this linear system. In addition, we will ensure that also the behavior $||\mathcal{A}||$ of the weighted automaton itself will constitute a component of the solution. To this end, the following identities will be very helpful.

4.3.19 Lemma. *Let $\mathcal{A} = (Q, \text{in}, \text{wt}, \text{out})$ be a weighted automaton over Σ with weights in S. Then its behavior satisfies the identity*

$$||\mathcal{A}|| = \sum_{(p,a,q) \in Q \times \Sigma \times Q} \left(\text{in}(p)\, \text{wt}(p,a,q)\, a \cdot ||\mathcal{A}_q|| \right) + \sum_{q \in Q} \text{in}(q)\, \text{out}(q)\varepsilon. \tag{4}$$

In particular, we have

$$||\mathcal{A}_p|| = \sum_{(a,q) \in \Sigma \times Q} \left(\text{wt}(p,a,q)\, a \cdot ||\mathcal{A}_q|| \right) + \text{out}(p)\varepsilon. \tag{5}$$

for any state $p \in Q$.

Proof. Considering the empty word, any path P in $\mathcal{A}$ with label ε is of the form $P = q$ for some $q \in Q$. Thus, we obtain

$$(||\mathcal{A}||, \varepsilon) = \sum_{\substack{P \in \text{Path}(\mathcal{A}) \\ \text{label}(P) = \varepsilon}} \text{weight}(P) = \sum_{q \in Q} \text{in}(q)\, \text{out}(q)$$

and by definition we have $(\text{in}(p)\, \text{wt}(p,a,q)a, \varepsilon) = 0$ for any transition (p,a,q). Consequently, identity (4) holds true for the empty word.

On the other hand, given a non-empty word $w \in \Sigma^+$, we have $(\text{in}(q)\,\text{out}(q)\varepsilon, w) = 0$ for any $q \in Q$, and hence it suffices to show

$$(||\mathcal{A}||, w) = \sum_{(p,a,q)\in Q\times\Sigma\times Q} \big(\text{in}(p)\,\text{wt}(p,a,q)a \cdot ||\mathcal{A}_q||, w\big).$$

We first observe that we have $\text{Path}(\mathcal{A}_q) = \text{Path}(\mathcal{A})$ for any state $p \in Q$, since the weighted automata $\mathcal{A}_p$ and $\mathcal{A}$ have the same state set. If we now write $w = a'w'$ for suitable $a' \in \Sigma$ and $w' \in \Sigma^*$, then any path P with label w has the form $P = pa'P'$, where p is a state and P' is a path with label w'. Assuming that the path P' starts in state q and leads to state r, we further obtain

$$\begin{aligned}
\text{weight}_{\mathcal{A}}(P) &= \text{in}(p)\,\text{wt}(p,a',q) \cdot \text{rweight}(P') \cdot \text{out}(r) \\
&= \text{in}(p)\,\text{wt}(p,a',q) \cdot \text{in}_q(q) \cdot \text{rweight}(P') \cdot \text{out}(r) \\
&= \text{in}(p)\,\text{wt}(p,a',q) \cdot \text{weight}_{\mathcal{A}_q}(P'),
\end{aligned}$$

since by definition $\text{in}_q(q) = 1$. On the other hand, we have $\text{in}_q(q') = 0$ for any $q' \neq q$, and thus, assuming that P' starts in q', also $\text{weight}_{\mathcal{A}_q}(P') = 0$. Exploiting these observations, we compute

$$\begin{aligned}
(||\mathcal{A}||, w) &= \sum_{\substack{P\in\text{Path}(\mathcal{A}) \\ \text{label}(P)=w}} \text{weight}_{\mathcal{A}}(P) \\
&= \sum_{p,q\in Q}\sum_{r\in Q}\sum_{P':\,q\xrightarrow{w'}r} \text{in}(p)\,\text{wt}(p,a',q) \cdot \text{weight}_{\mathcal{A}_q}(P') \\
&= \sum_{p,q\in Q} \text{in}(p)\,\text{wt}(p,a',q) \cdot \sum_{r\in Q}\sum_{P':\,q\xrightarrow{w'}r} \text{weight}_{\mathcal{A}_q}(P') \\
&= \sum_{p,q\in Q} \text{in}(p)\,\text{wt}(p,a',q) \cdot \sum_{\substack{P'\in\text{Path}(\mathcal{A}) \\ \text{label}(P')=w'}} \text{weight}_{\mathcal{A}_q}(P') \\
&= \sum_{p,q\in Q} \text{in}(p)\,\text{wt}(p,a',q) \cdot (||\mathcal{A}_q||, w') \\
&= \sum_{(p,a,q)\in Q\times\Sigma\times Q}\sum_{w=uv} \underbrace{(\text{in}(p)\,\text{wt}(p,a,q)a, u)}_{=0 \text{ if } u\neq a} \cdot (||\mathcal{A}_q||, v) \\
&= \sum_{(p,a,q)\in Q\times\Sigma\times Q} \big(\text{in}(p)\,\text{wt}(p,a,q)a \cdot ||\mathcal{A}_q||, w\big).
\end{aligned}$$

This establishes identity (4). Given any state $p \in Q$ and applying identity (4) to the behavior of the automaton $\mathcal{A}_p$, we obtain identity (5), since by definition we have

$$\text{in}_p(q) = \begin{cases} 1 & \text{if } q = p \\ 0 & \text{otherwise.} \end{cases}$$

$\square$

Using the identities provided in Lemma 4.3.19, we are now ready to transform a weighted automaton into a proper linear system with rational coefficients.

4.3.20 Proposition. *Let $r \in S^{\text{rec}}\langle\!\langle \Sigma^* \rangle\!\rangle$ be a recognizable formal power series. Then there exist an $n \in \mathbb{N}$, rational power series $r_{ij}, r_i \in S^{\text{rat}}\langle\!\langle \Sigma^* \rangle\!\rangle$ such that r_{ij} is proper for any $i, j \in \{0, \ldots, n\}$, and a power series $z_i \in S\langle\!\langle \Sigma^* \rangle\!\rangle$ for any $i \in \{0, \ldots, n\}$ such that $(z_0, \ldots, z_n)$ is a solution of the proper linear system (2) and the component z_0 coincides with the recognizable power series r.*

Proof. Let $\mathcal{A} = (Q, \text{in}, \text{wt}, \text{out})$ be a weighted automaton such that $||\mathcal{A}|| = r$. As the state set is non-empty and finite, we can write it as $Q = \{q_1, \ldots, q_n\}$ for some $n \in \mathbb{N}^+$. For $i, j \in \{0, \ldots, n\}$ we define power series $r_{ij}, r_i \in S\langle\!\langle \Sigma^* \rangle\!\rangle$ by:

$$
r_{ij} = \begin{cases} \mathbb{1}_\emptyset = 0\varepsilon & \text{if } j = 0 \\ \sum_{p \in Q, a \in \Sigma} (\text{in}(p)\,\text{wt}(p, a, q_j))a & \text{if } i = 0 \text{ and } j \in \{1, \ldots, n\} \\ \sum_{a \in \Sigma} \text{wt}(q_i, a, q_j)a & \text{if } i, j \in \{1, \ldots, n\} \end{cases}
$$

$$
r_i = \begin{cases} \sum_{q \in Q} \text{in}(q)\,\text{out}(q)\varepsilon & \text{if } i = 0 \\ \text{out}(q_i)\varepsilon & \text{if } i \in \{1, \ldots, n\} \end{cases}
$$

For any $i, j \in \{0, \ldots, n\}$ the power series r_{ij} and r_i are polynomials and hence rational (see Remark 4.2.2). Moreover, the power series r_{ij} is proper, since $\varepsilon \notin \text{supp}(r_{ij})$. If we further set $z_0 = ||\mathcal{A}||$ and $z_i = ||\mathcal{A}_{q_i}||$ for any $i \in \{1, \ldots, n\}$, then the identities from Lemma 4.3.19 and the distributivity in $S\langle\!\langle \Sigma^* \rangle\!\rangle$ imply that $(z_0, \ldots, z_n)$ is a solution of the proper linear system

$$
\left(X_i = \sum_{j=0}^{n} r_{ij} X_j + r_i \right)_{0 \leq i \leq n} .
$$

As a final observation, by definition the component z_0 coincides with the recognizable power series $||\mathcal{A}|| = r$. $\qquad\square$

Proof of Theorem 4.3.1. In order to prove the inclusion $S^{\text{rat}}\langle\!\langle \Sigma^* \rangle\!\rangle \subseteq S^{\text{rec}}\langle\!\langle \Sigma^* \rangle\!\rangle$, we proceed by structural induction using Proposition 4.3.2, Proposition 4.3.3, Proposition 4.3.5 and Proposition 4.3.6. The other inclusion $S^{\text{rec}}\langle\!\langle \Sigma^* \rangle\!\rangle \subseteq S^{\text{rat}}\langle\!\langle \Sigma^* \rangle\!\rangle$ follows by Proposition 4.3.20 and Proposition 4.3.17. More precisely, by Proposition 4.3.20 any recognizable power series r is a component of the solution vector of a proper linear system with rational components. Hence, Proposition 4.3.17 yields that the recognizable power series r must also be rational. $\qquad\square$

To prove the inclusion $S^{\text{rat}}\langle\!\langle \Sigma^* \rangle\!\rangle \subseteq S^{\text{rec}}\langle\!\langle \Sigma^* \rangle\!\rangle$ we have proceeded by structural induction and exploited the expressive equivalence of weighted automata and linear representations (see Proposition 3.3.8). As we have already mentioned at the end of Section 3.3, any weighted automaton $\mathcal{A}$ can effectively be transformed into a linear representation (λ, μ, γ) such that $||(\lambda, \mu, \gamma)|| = ||\mathcal{A}||$ and viceversa (provided the underlying semiring is *computable*). Thus, we obtain an effective procedure that, given a rational power series (or to be precise: a *weighted rational expression*), inductively computes a weighted automaton $\mathcal{A}$ such that $||\mathcal{A}|| = r$. Indeed, Proposition 4.3.2, Proposition 4.3.3, Proposition 4.3.5 and Proposition 4.3.6 all come with effective procedures. Conversely, the proof of Proposition 4.3.20 provides an effective procedure that, given a weighted automaton $\mathcal{A}$, produces a proper linear system whose coefficients are polynomials and thus rational. Following the proof of Proposition 4.3.17, we can then proceed inductively to construct $||\mathcal{A}||$ from these polynomials by finitely many applications of the rational operations sum, Cauchy product and Kleene star, which yields that $||\mathcal{A}||$ is rational, i.e. $||\mathcal{A}|| \in S^{\mathrm{rat}}\langle\!\langle \Sigma^* \rangle\!\rangle$.

4.3.21 Example. Consider the alphabet $\Sigma = \{x\}$ and the semiring $\mathbb{N}$ of natural numbers. We recall from Example 3.2.12 that the weighted automaton $\mathcal{A}$ given by

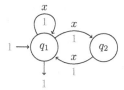

has as behavior the power series

$$||\mathcal{A}|| = \sum_{n \in \mathbb{N}} a_n x^n,$$

whose coefficients form the Fibonacci sequence. More precisely, we have $a_0 = a_1 = 1$ and $a_{n+2} = a_{n+1} + a_n$ for $n \in \mathbb{N}$. Following the proof of Proposition 4.3.20, we define rational power series, which are even polynomials, by

$r_{00} = \mathbb{1}_\emptyset$	$r_{01} = 1x$	$r_{02} = 1x$	$r_0 = 1\varepsilon$
$r_{10} = \mathbb{1}_\emptyset$	$r_{11} = 1x$	$r_{12} = 1x$	$r_1 = 1\varepsilon$
$r_{20} = \mathbb{1}_\emptyset$	$r_{21} = 1x$	$r_{22} = 0x$	$r_2 = 0\varepsilon$

to obtain the following proper linear system:

$$X_0 = 1xX_1 + 1xX_2 + 1\varepsilon$$
$$X_1 = 1xX_1 + 1xX_2 + 1\varepsilon$$
$$X_2 = 1xX_1$$

The first two equations imply the equation $X_0 = X_1$. Thus, by substitution the system can be reduced to the single equation

$$X_0 = 1xX_0 + 1x^2X_0 + 1\varepsilon,$$

since $(1x)^2 = 1x^2$.

115

By Lemma 4.3.14 the unique solution z of this proper linear equation is given by the power series

$$z = \left(1x + 1x^2\right)^* 1\varepsilon = \left(1x + 1x^2\right)^*.$$

On the other hand, the definition of the above system guarantees that also $\|\mathcal{A}\|$ solves the equation $X_0 = 1xX_0 + 1x^2X_0 + 1\varepsilon$ (see Proposition 4.3.20). Thus, the uniqueness of the solution z yields

$$\|\mathcal{A}\| = \left(1x + 1x^2\right)^*,$$

which in particular implies that

$$\|\mathcal{A}\| = \sum_{n \in \mathbb{N}} a_n x^n$$

is a rational power series.

The procedure provided by Proposition 4.3.17 and Proposition 4.3.20 can in particular be performed for the Boolean semiring $\mathbb{B}$ to show that every recognizable language is rational. Perrin [27, page 16] directly performs this approach for languages, without the detour involving power series.

4.3.22 Remark. Any map from Σ^* into the multiplicative monoid $S^{Q \times Q}$ of matrices can alternatively be regarded as a matrix (of dimension Q) whose entries are power series in $S\langle\!\langle \Sigma^* \rangle\!\rangle$ (cf. Droste and Kuich [9, page 19]). Thus, any linear representation (λ, μ, γ) provides coefficients for a linear system. More precisely, the monoid homomorphism $\mu \colon \Sigma^* \to S^{Q \times Q}$ can be regarded as matrix in $(S\langle\!\langle \Sigma^* \rangle\!\rangle)^{Q \times Q}$, which then can be used as coefficient matrix of a linear system (see Remark 4.3.16). This draws a further connection between recognizable power series and linear systems. We refer to Droste, Kuich and Vogler [8, Chapter 3], Kuich and Salomaa [21, § II.7], Salomaa and Soittola [35, § II.1 and § II.2] and Eilenberg [14, Chapter VII, § 6] for further details on linear systems in the light of weighted automata and formal power series. For instance, [8, Chapter 3, Theorem 2.18] provides a characterization of recognizable power series by means of linear systems, which in a sense extends Proposition 4.3.20. Moreover, in Berstel and Reutenauer [2, Chapter I, Theorem 6.1], Droste [11, Theorem 6.19], [35, Chapter II, Theorem 2.3] and [14, Chapter VII, Theorem 5.1] one finds alternative proofs for Theorem 4.3.1.

4.3.23 Remark. We have seen above that the proofs of both directions in Theorem 4.3.1 come with effective procedures. Thus, all decidability and undecidability results regarding weighted automata and their behavior transfer also to rational power series and viceversa. In [2, Chapter VI], Berstel and Reutenauer present several decidability results for rational power series. In particular, they derive fundamental properties concerning supports, like e.g. the decidability of the Emptiness Problem and the Finiteness Problem. Moreover, they show that it is decidable whether a rational power series has finite image. Recall that this problem is of particular interest, as the support of any rational power series with finite image is a rational language, if the underlying semiring is a commutative ring (see Remark 4.2.5). In [12, § 8], Droste and Kuske study the decidability of the Equivalence Problem for recognizable power series, i.e. the question whether two given recognizable power series are equal, which in general is not decidable. Further decidability and undecidability results regarding recognizable and rational power series over the ring $\mathbb{Z}$ of integers can be found in Salomaa and Soittola [35, § II.12].

Due to Theorem 4.3.1, there is no difference between the notions of recognizability and rationality for power series. In particular, all properties of recognizable power series transfer to rational power series and viceversa. Berstel and Reutenauer use the equivalence provided by Theorem 4.3.1 as starting point for their developments in [2]. In [35, Chapter III], Salomaa and Soittola treat several applications of rational and recognizable power series, like e.g. *rational transductions*.

Having derived a generalization of Kleene's Theorem, we next aim at similarly generalizing the Büchi–Elgot–Trakhtenbrot Theorem. To this end, we introduce a weighted version of monadic second-order logic in the context of formal power series, which is due to Droste and Gastin [5]. Employing sentences of this weighted monadic second-order logic, we then obtain a further formalism for the modeling of formal power series. We will show that, under particular assumptions, certain restrictions of this formalism have the same expressive power as weighted automata.

5. Weighted Monadic Second-Order Logic and Weighted Automata

With their fundamental result, Büchi [4], Elgot [15] and Trakhtenbrot [43] established the expressive equivalence of classical automata and monadic second-order logic. At the same time, Schützenberger [37] investigated formal power series in the context of Automata Theory, introduced the notion of weighted automata, and characterized their behaviors as rational formal power series. Hence, he established a generalization of Kleene's Theorem, which we have presented in Chapter 4. In 2005, Droste and Gastin [5] extended the Büchi–Elgot–Trakhtenbrot Theorem to the realm of formal power series. They did so by presenting a weighted version of monadic second-order logic, which is capable of determining quantitative properties of words instead of just qualitative truth statements. Moreover, they analyzed conditions under which the behaviors of weighted automata are precisely the power series which are definable by some sentence of their weighted monadic second-order logic. As intended, Droste and Gastin could show that, under certain assumptions on the underlying semiring, particular restrictions of their weighted logic are expressively equivalent to weighted automata. Thus, their work gives rise to a generalization of the classical result of Büchi, Elgot and Trakhtenbrot.

The goal of this chapter is to give an introduction to the work of Droste and Gastin. In Section 5.1, we define their quantitative extension of monadic second-order logic. We draw connections to classical monadic second-order logic and present various examples to illustrate the expressive power of weighted formulas and to point out possible application scenarios. Some of these examples exhibit that the formalism of weighted monadic second-order formulas is more expressive than weighted automata. Therefore, we ultimately introduce several fragments and restrictions of the weighted logic, respectively. In Section 5.2, we aim at deriving the results of Droste and Gastin. More precisely, we carry out their proofs by outlining the main steps of Droste and Gastin [6], which is a more detailed version of their work [5] from 2005. We first establish the expressive equivalence between weighted automata and restricted sentences of weighted monadic second-order logic for commutative semirings. Secondly, we introduce the notion of locally finite semirings, and we show that for this large class of semirings also unrestricted sentences realize recognizable power series.

We recall that throughout this work Σ denotes an alphabet and S denotes a non-trivial semiring.

5.1. Syntax and Semantics

In [5], Droste and Gastin's aim is to introduce a logic that forms a quantitative extension of monadic second-order logic for words just as weighted automata extend the classical ones. To this end, they incorporate weights in monadic second-order logic. More precisely, they enrich the syntax of the classical logic $\mathrm{MSO}(\Sigma)$ by permitting all elements of a given semiring S as formulas. As the behavior of a weighted automaton is a formal power series, the semantics of a weighted sentence should as well be a formal power series. More generally, they define the semantics of a weighted formula (which may have free variables) by structural induction as a formal power series over an extended alphabet. Indeed, they are able to assign natural semantics to atomic formulas, to disjunctions and conjunctions, as well as to existential and universal quantifications. However, with the negation of formulas there arises a problem. As semirings do in general not have a complement operation, the semantics of negated formulas cannot be defined elementwise. Therefore, negation is restricted to atomic formulas, whose semantics will be unambiguous power series, i.e. will only take the values 0 and 1. In comparison to the classical logic $\mathrm{MSO}(\Sigma)$, this is not an essential restriction, as conjunction and universal quantification are included into the syntax of weighted monadic second-order logic. Hence, the classical logic $\mathrm{MSO}(\Sigma)$ is contained in the weighted logic. For the Boolean semiring, the two formalisms do even coincide. However, e.g. for the semiring of natural numbers, it turns out that the semantics of a weighted formula is in general not a recognizable power series, as universal quantifications do not preserve recognizability. Consequently, one has to consider fragments of the weighted logic in order to obtain a formalism that is expressively equivalent to weighted automata. To this end, Droste and Gastin introduce a restricted fragment of their weighted logic by excluding second-order universal quantification and permitting first-order universal quantification only for formulas whose semantics take just finitely many values.

The main objective of this section is to formally introduce the syntax and semantics of the weighted monadic second-order logic of Droste and Gastin. Furthermore, we draw connections to the classical monadic second-order logic $\mathrm{MSO}(\Sigma)$ for words, as studied in see Section 2.4, by considering the Boolean semiring. We also consider other semirings and give examples of particular weighted formulas and their semantics to illustrate the expressive power and possible applications scenarios of the weighted monadic second-order logic. In particular, we present weighted sentences whose semantics are not recognizable power series. Therefore, we ultimately introduce the fragment of *restricted* weighted formulas, which we will use in Section 5.2 to characterize the behaviors of weighted automata. We set up notation and terminology following Droste and Gastin [6, § 3]. However, we also took inspiration from Droste and Gastin [10, § 3] as well as from Droste and Kuske [12, § 7].

As in Section 2.4, we consider the relational signature $\sigma_\Sigma = (\leq, (P_a)_{a \in \Sigma})$, where $\leq$ denotes a binary relation symbol and P_a a unary relation symbol for every letter $a \in \Sigma$. Moreover, we fix two countably infinite disjoint sets Var_1 and Var_2 of **first-order** and **second-order variables**, respectively, and we set $\mathrm{Var} := \mathrm{Var}_1 \cup \mathrm{Var}_2$.

120

In the following, we refer to $\text{MSO}(\Sigma)$ as *classical* monadic second-order logic to stress the difference to the weighted logic, which we will introduce in this section. We now define the formal syntax of the weighted version of monadic second-order logic.

5.1.1 Definition. We define the set of **weighted monadic second-order formulas** over Σ and S, or the set of (weighted) $\text{MSO}(\Sigma, S)$–formulas, inductively as follows:

(1) s is an $\text{MSO}(\Sigma, S)$–formula for any $s \in S$.

(2) $P_a(x)$ is an $\text{MSO}(\Sigma, S)$–formula for each $a \in \Sigma$ and $x \in \text{Var}_1$.

(3) $x \le y$ is an $\text{MSO}(\Sigma, S)$–formula for any $x, y \in \text{Var}_1$.

(4) $x \in X$ is an $\text{MSO}(\Sigma, S)$–formula for each $x \in \text{Var}_1$ and $X \in \text{Var}_2$.

We call the formulas obtained by steps (2)–(4) **atomic**.

(5) The **negation** $\neg\varphi$ of each atomic $\text{MSO}(\Sigma, S)$–formula is an $\text{MSO}(\Sigma, S)$–formula.

(6) If φ and ψ are $\text{MSO}(\Sigma, S)$–formulas, then so are their **disjunction** $(\varphi \vee \psi)$ and **conjunction** $(\varphi \wedge \psi)$.

(7) If φ is an $\text{MSO}(\Sigma, S)$–formula and $x \in \text{Var}_1$, then the **first-order existential quantification** $\exists x\, \varphi$ and the **first-order universal quantification** $\forall x\, \varphi$ are again $\text{MSO}(\Sigma, S)$–formulas.

(8) If φ is an $\text{MSO}(\Sigma, S)$–formula and $X \in \text{Var}_2$, then the **second-order existential quantification** $\exists X\varphi$ and the **second-order universal quantification** $\forall X\varphi$ are again $\text{MSO}(\Sigma, S)$–formulas.

We write $\varphi \in \text{MSO}(\Sigma, S)$ if φ is an $\text{MSO}(\Sigma, S)$–formula. The formulas obtained by steps (2)–(5) are called **literals**.

5.1.2 Remark.

a) First, we observe that in $\text{MSO}(\Sigma, S)$ negation is restricted to atomic formulas. In fact, the negation of general $\text{MSO}(\Sigma, S)$–formulas is not permitted due to difficulties defining then their semantics. As we have already mentioned, the semantics of a weighted formula $\varphi \in \text{MSO}(\Sigma, S)$ will be represented by a formal power series over an (extended) alphabet with coefficients in S. It would then be natural to define the power series representing the semantics of its negation $\neg\varphi$ pointwise. In fact, this is possible if the underlying semiring S is induced by a *bounded distributive lattice* with complement function, like e.g. any Boolean algebra (see Remark 5.1.6), or the semiring $([0, 1], \max, \min, 0, 1)$ with complement function $x \mapsto 1 - x$ for $x \in [0, 1]$ (cf. [10, page 181] and Rahonis [32, §3.3]). However, semirings do in general not have a natural complement function. Therefore, negation is restricted to atomic formulas, whose semantics will constitute unambiguous formal power series. Thus, the negation of any atomic formula has a natural, pointwise defined semantics, since unambiguous power series take only the values 0 and 1, which act complementary. Due to this restriction of negation to atomic formulas, conjunction and universal quantifications are included into the syntax of the weighted logic $\text{MSO}(\Sigma, S)$.

b) In comparison to the classical logic $\mathrm{MSO}(\Sigma)$, the restriction of negation is not an essential restriction. Indeed, if we also include conjunction and universal quantifications into the syntax of $\mathrm{MSO}(\Sigma)$, then the negation of a classical $\mathrm{MSO}(\Sigma)$–formula is equivalent (in the sense of defining the same language) to one in which negation is applied only to atomic formulas (see Remark A.1.20 and Remark A.2.8). Thus, for any $\varphi \in \mathrm{MSO}(\Sigma)$ there exists a $\psi \in \mathrm{MSO}(\Sigma)$ (possibly containing conjunctions and universal quantifications) in which negation is only applied to atomic formulas and which fulfills $L(\psi) = L(\varphi)$. This classical $\mathrm{MSO}(\Sigma)$–formula ψ can be viewed as weighted $\mathrm{MSO}(\Sigma, S)$–formula for any semiring S. In this sense, the weighted logic $\mathrm{MSO}(\Sigma, S)$ contains the classical logic $\mathrm{MSO}(\Sigma)$ for any semiring S.

c) The main difference to the syntax of the classical logic $\mathrm{MSO}(\Sigma)$ is that $\mathrm{MSO}(\Sigma, S)$ incorporates weights taken from the semiring S. More precisely, each weight $s \in S$ is realized by the $\mathrm{MSO}(\Sigma, S)$–formula $\varphi = s$. If we consider the Boolean semiring $\mathbb{B}$, then the only weights that may occur in an $\mathrm{MSO}(\Sigma, \mathbb{B})$–formula are 0 and 1. These correspond to the classical $\mathrm{MSO}(\Sigma)$–formulas $\bot$ and $\top$ (see Remark A.2.6). In particular, the weighted logic $\mathrm{MSO}(\Sigma, \mathbb{B})$ corresponds precisely to the classical logic $\mathrm{MSO}(\Sigma)$ (see Proposition 5.1.10).

5.1.3 Definition. Given $\varphi \in \mathrm{MSO}(\Sigma, S)$, we define the set $\mathrm{Free}(\varphi)$ of **free variables** of the weighted formula φ by structural induction as follows:

$$\mathrm{Free}(s) := \emptyset,$$
$$\mathrm{Free}(P_a(x)) := \{x\},$$
$$\mathrm{Free}(x \leq y) := \{x, y\},$$
$$\mathrm{Free}(x \in X) := \{x, X\},$$
$$\mathrm{Free}(\neg \varphi) := \mathrm{Free}(\varphi),$$
$$\mathrm{Free}(\varphi \vee \psi) := \mathrm{Free}(\varphi \wedge \psi) := (\mathrm{Free}(\varphi) \cup \mathrm{Free}(\psi)),$$
$$\mathrm{Free}(\exists x\, \varphi) := \mathrm{Free}(\forall x\, \varphi) := \mathrm{Free}(\varphi) \setminus \{x\},$$
$$\mathrm{Free}(\exists X \varphi) := \mathrm{Free}(\forall X \varphi) := \mathrm{Free}(\varphi) \setminus \{X\}.$$

A formula $\varphi \in \mathrm{MSO}(\Sigma, S)$ with $\mathrm{Free}(\varphi) = \emptyset$ is called an $\mathrm{MSO}(\Sigma, S)$–**sentence**.

To define the semantics of the weighted logic $\mathrm{MSO}(\Sigma, S)$, we first recall some notions in connection with the semantics of the classical logic $\mathrm{MSO}(\Sigma)$. Each word w over Σ provides a σ_Σ–structure $(\mathrm{dom}(w), \leq, (P_a)_{a \in \Sigma})$, where $\mathrm{dom}(w) = \{1, \ldots, |w|\} \subseteq \mathbb{N}$ is the set of positions in w and $P_a = P_a^w = \{i \in \mathrm{dom}(w) \mid w(i) = a\}$ is the interpretation of the unary relation symbol P_a for any $a \in \Sigma$ (see Definition 2.4.4). Furthermore, given a finite set $\mathcal{V} \subseteq \mathrm{Var}$ of variables and a word $w \in \Sigma^*$, a $(\mathcal{V}, w)$–assignment is a map $\sigma : \mathcal{V} \mapsto \mathrm{dom}(w) \cup \mathcal{P}(\mathrm{dom}(w))$ that assigns positions to first-order variables in $\mathcal{V}$ and sets of positions in w to second-order variables in $\mathcal{V}$ (see Definition 2.4.8). We encode pairs (w, σ) consisting of a word w over Σ and a $(\mathcal{V}, w)$–assignment σ as words over the extended alphabet $\Sigma_\mathcal{V} = \Sigma \times \{0, 1\}^\mathcal{V}$ and we denote by $N_\mathcal{V}$ the set that comprises all words over $\Sigma_\mathcal{V}$ that correspond to such a pair (w, σ) (see Definition 2.5.3).

Given a classical MSO(Σ)–formula φ and a finite set $\mathcal{V}$ of variables containing Free(φ), the language $L_\mathcal{V}(\varphi)$ defined by φ consists of all words $v = (w, \sigma) \in N_\mathcal{V}$ that satisfy φ, i.e. that fulfill $(w, \sigma) \models \varphi$ (see Definition 2.4.9 and Definition 2.5.6).

5.1.4 Definition. Let $\varphi \in \mathrm{MSO}(\Sigma, S)$ be a weighted formula and $\mathcal{V} \subseteq \mathrm{Var}$ a finite set of variables such that Free(φ) $\subseteq \mathcal{V}$. The $\mathcal{V}$–**semantics** of φ is a formal power series

$$\llbracket \varphi \rrbracket_\mathcal{V} \in S\langle\langle \Sigma_\mathcal{V}^* \rangle\rangle.$$

Given $v \in \Sigma_\mathcal{V}^*$, to define the coefficient $(\llbracket \varphi \rrbracket_\mathcal{V}, v) \in S$ we distinguish two cases:

- If $v \in \Sigma_\mathcal{V}^*$ is a non-valid word, i.e. $v \notin N_\mathcal{V}$, then we set $(\llbracket \varphi \rrbracket_\mathcal{V}, v) = 0$.

- If $v \in \Sigma_\mathcal{V}^*$ is a valid word, i.e. $v = (w, \sigma) \in N_\mathcal{V}$, then either $v = (w, \sigma) = \varepsilon$ or the pair (w, σ) consists of a non-empty word $w \in \Sigma^+$ and a $(\mathcal{V}, w)$–assignment σ. In both cases, we define the coefficient

$$(\llbracket \varphi \rrbracket_\mathcal{V}, v) := \llbracket \varphi \rrbracket_\mathcal{V}(w, \sigma)$$

by structural induction as follows:

$$\llbracket s \rrbracket_\mathcal{V}(w, \sigma) := s,$$

$$\llbracket P_a(x) \rrbracket_\mathcal{V}(w, \sigma) := \begin{cases} 1 & \text{if } (w, \sigma) \models P_a(x) \\ 0 & \text{otherwise}, \end{cases}$$

$$\llbracket x \le y \rrbracket_\mathcal{V}(w, \sigma) := \begin{cases} 1 & \text{if } (w, \sigma) \models x \le y \\ 0 & \text{otherwise}, \end{cases}$$

$$\llbracket x \in X \rrbracket_\mathcal{V}(w, \sigma) := \begin{cases} 1 & \text{if } (w, \sigma) \models x \in X \\ 0 & \text{otherwise}, \end{cases}$$

$$\llbracket \neg \varphi \rrbracket_\mathcal{V}(w, \sigma) := \begin{cases} 1 & \text{if } \llbracket \varphi \rrbracket_\mathcal{V}(w, \sigma) = 0 \\ 0 & \text{if } \llbracket \varphi \rrbracket_\mathcal{V}(w, \sigma) = 1 \end{cases} \quad (\text{if } \varphi \text{ is atomic}),$$

$$\llbracket \varphi \vee \psi \rrbracket_\mathcal{V}(w, \sigma) := \llbracket \varphi \rrbracket_\mathcal{V}(w, \sigma) + \llbracket \psi \rrbracket_\mathcal{V}(w, \sigma),$$

$$\llbracket \varphi \wedge \psi \rrbracket_\mathcal{V}(w, \sigma) := \llbracket \varphi \rrbracket_\mathcal{V}(w, \sigma) \cdot \llbracket \psi \rrbracket_\mathcal{V}(w, \sigma),$$

$$\llbracket \exists x\, \varphi \rrbracket_\mathcal{V}(w, \sigma) := \sum_{i \in \mathrm{dom}(w)} \llbracket \varphi \rrbracket_{\mathcal{V} \cup \{x\}}(w, \sigma[x \mapsto i]),$$

$$\llbracket \forall x\, \varphi \rrbracket_\mathcal{V}(w, \sigma) := \prod_{i \in \mathrm{dom}(w)} \llbracket \varphi \rrbracket_{\mathcal{V} \cup \{x\}}(w, \sigma[x \mapsto i]),$$

$$\llbracket \exists X \varphi \rrbracket_\mathcal{V}(w, \sigma) := \sum_{I \subseteq \mathrm{dom}(w)} \llbracket \varphi \rrbracket_{\mathcal{V} \cup \{X\}}(w, \sigma[X \mapsto I]),$$

$$\llbracket \forall X \varphi \rrbracket_\mathcal{V}(w, \sigma) := \prod_{I \subseteq \mathrm{dom}(w)} \llbracket \varphi \rrbracket_{\mathcal{V} \cup \{X\}}(w, \sigma[X \mapsto I]).$$

We calculate the products in the coefficients of $[\![\forall x\, \varphi]\!]_{\mathcal{V}}$ and $[\![\forall X \varphi]\!]_{\mathcal{V}}$ following the natural enumeration of $\mathrm{dom}(w) = \{1, \ldots, |w|\}$ and some fixed enumeration of its power set $\mathcal{P}(\mathrm{dom}(w))$, respectively (see Remark 5.1.7).

We simply write $[\![\varphi]\!]$ for the $\mathrm{Free}(\varphi)$–semantics $[\![\varphi]\!]_{\mathrm{Free}(\varphi)} \in S\langle\!\langle \Sigma_\varphi^* \rangle\!\rangle$. Moreover, we usually refer to the $\mathrm{Free}(\varphi)$–semantics $[\![\varphi]\!]$ of a weighted $\mathrm{MSO}(\Sigma, S)$–formula φ simply as its semantics.

5.1.5 Remark.

a) For any $s \in S$, the semantics of the weighted $\mathrm{MSO}(\Sigma, S)$–sentence s is given by the constant power series

$$[\![s]\!] = \sum_{w \in \Sigma^*} sw = s \cdot \mathbf{1}_{\Sigma^*} \in S\langle\!\langle \Sigma^* \rangle\!\rangle.$$

More generally, we have $[\![s]\!]_{\mathcal{V}} = s \cdot \mathbf{1}_{N_{\mathcal{V}}} \in S\langle\!\langle \Sigma_{\mathcal{V}}^* \rangle\!\rangle$.

b) The $\mathcal{V}$–semantics of any literal $\varphi \in \mathrm{MSO}(\Sigma, S)$ is an unambiguous formal power series in $S\langle\!\langle \Sigma_{\mathcal{V}}^* \rangle\!\rangle$, i.e. $[\![\varphi]\!]_{\mathcal{V}}$ takes only the values 0 and 1. More precisely, if we consider φ also as classical $\mathrm{MSO}(\Sigma)$–formula, then we have

$$[\![\varphi]\!]_{\mathcal{V}} = \mathbf{1}_{L_{\mathcal{V}}(\varphi)} \text{ and } \mathrm{supp}([\![\varphi]\!]_{\mathcal{V}}) = L_{\mathcal{V}}(\varphi),$$

where $L_{\mathcal{V}}(\varphi) = \{(w, \sigma) \in N_{\mathcal{V}} \mid (w, \sigma) \models \varphi\}$. Indeed, e.g. for the literal $\neg P_a(x)$, we achieve

$$([\![\neg P_a(x)]\!]_{\mathcal{V}}, v) \neq 0 \;\Leftrightarrow\; ([\![\neg P_a(x)]\!]_{\mathcal{V}}, v) = 1$$
$$\Leftrightarrow\; v = (w, \sigma) \in N_{\mathcal{V}} \text{ and } [\![\neg P_a(x)]\!]_{\mathcal{V}}(w, \sigma) = 1$$
$$\Leftrightarrow\; v = (w, \sigma) \in N_{\mathcal{V}} \text{ and } [\![P_a(x)]\!]_{\mathcal{V}}(w, \sigma) = 0$$
$$\Leftrightarrow\; v = (w, \sigma) \in N_{\mathcal{V}} \text{ and } (w, \sigma) \not\models P_a(x)$$
$$\Leftrightarrow\; v = (w, \sigma) \in N_{\mathcal{V}} \text{ and } (w, \sigma) \models \neg P_a(x)$$
$$\Leftrightarrow\; v = (w, \sigma) \in L_{\mathcal{V}}(\neg P_a(x))$$

for any word v over the extended alphabet $\Sigma_{\mathcal{V}}$.

c) The $\mathcal{V}$–semantics of a disjunction $\varphi \vee \psi$ is obtained by pointwise adding the coefficients of $[\![\varphi]\!]_{\mathcal{V}}$ and $[\![\psi]\!]_{\mathcal{V}}$, i.e. we have

$$[\![\varphi \vee \psi]\!]_{\mathcal{V}} = [\![\varphi]\!]_{\mathcal{V}} + [\![\psi]\!]_{\mathcal{V}}.$$

Similarly, one obtains that the $\mathcal{V}$–semantics of a conjunction $\varphi \wedge \psi$ is given by the Hadamard product

$$[\![\varphi \wedge \psi]\!]_{\mathcal{V}} = [\![\varphi]\!]_{\mathcal{V}} \odot [\![\psi]\!]_{\mathcal{V}}.$$

As the underlying semiring S is not assumed to be commutative, the semantics of conjunction does in general not constitute a commutative operation, i.e. we have

$$[\![\varphi \wedge \psi]\!]_{\mathcal{V}} = [\![\varphi]\!]_{\mathcal{V}} \odot [\![\psi]\!]_{\mathcal{V}} \neq [\![\psi]\!]_{\mathcal{V}} \odot [\![\varphi]\!]_{\mathcal{V}} = [\![\psi \wedge \varphi]\!]_{\mathcal{V}}.$$

In contrast, conjunction in the classical logic $\mathrm{MSO}(\Sigma)$ does provide a commutative operation on languages.

124

d) We note further that, given a valid word $v \in \Sigma_{\mathcal{V}}^*$, i.e. $v \in N_{\mathcal{V}}$, its decoding (w, σ), a first-order variable x and a position $i \in \mathrm{dom}(w)$, the encoding v' of the pair $(w, \sigma[x \mapsto i])$ is again a valid word. More precisely we have

$$v' = (w, \sigma[x \mapsto i]) \in N_{\mathcal{V} \cup \{x\}} \subseteq \Sigma_{\mathcal{V} \cup \{x\}}^*.$$

Similarly, we achieve

$$(w, \sigma[X \mapsto I]) \in N_{\mathcal{V} \cup \{X\}} \subseteq \Sigma_{\mathcal{V} \cup \{X\}}^*$$

for any second-order variable X and any set $I \subseteq \mathrm{dom}(w)$ of positions.

e) If we consider the empty word, which belongs to $N_{\mathcal{V}}$, then we obtain

$$\llbracket s \rrbracket_{\mathcal{V}}(\varepsilon) = s,$$
$$\llbracket \varphi \rrbracket_{\mathcal{V}}(\varepsilon) = 1 \quad (\text{if } \varphi \text{ is atomic}),$$
$$\llbracket \neg\varphi \rrbracket_{\mathcal{V}}(\varepsilon) = 0 \quad (\text{if } \varphi \text{ is atomic}),$$
$$\llbracket \varphi \vee \psi \rrbracket_{\mathcal{V}}(\varepsilon) = \llbracket \varphi \rrbracket_{\mathcal{V}}(\varepsilon) + \llbracket \psi \rrbracket_{\mathcal{V}}(\varepsilon),$$
$$\llbracket \varphi \wedge \psi \rrbracket_{\mathcal{V}}(\varepsilon) = \llbracket \varphi \rrbracket_{\mathcal{V}}(\varepsilon) \cdot \llbracket \psi \rrbracket_{\mathcal{V}}(\varepsilon),$$
$$\llbracket \exists x\, \varphi \rrbracket_{\mathcal{V}}(\varepsilon) = \sum_{i \in \mathrm{dom}(\varepsilon) = \emptyset} \llbracket \varphi \rrbracket_{\mathcal{V} \cup \{x\}}(\varepsilon) = 0,$$
$$\llbracket \forall x\, \varphi \rrbracket_{\mathcal{V}}(\varepsilon) = \prod_{i \in \mathrm{dom}(\varepsilon) = \emptyset} \llbracket \varphi \rrbracket_{\mathcal{V} \cup \{x\}}(\varepsilon) = 1,$$
$$\llbracket \exists X \varphi \rrbracket_{\mathcal{V}}(\varepsilon) = \sum_{I \subseteq \mathrm{dom}(\varepsilon) = \emptyset} \llbracket \varphi \rrbracket_{\mathcal{V} \cup \{X\}}(\varepsilon) = \llbracket \varphi \rrbracket_{\mathcal{V} \cup \{X\}}(\varepsilon),$$
$$\llbracket \forall X \varphi \rrbracket_{\mathcal{V}}(\varepsilon) = \prod_{I \subseteq \mathrm{dom}(\varepsilon) = \emptyset} \llbracket \varphi \rrbracket_{\mathcal{V} \cup \{X\}}(\varepsilon) = \llbracket \varphi \rrbracket_{\mathcal{V} \cup \{X\}}(\varepsilon),$$

since empty sums have value 0 and empty products have value 1 (see Notation 3.1.3). We observe a strong similarity to the semantics for the empty word in the classical logic $\mathrm{MSO}(\Sigma)$. In particular, the $\mathcal{V}$–semantics for the empty word does not depend on the underlying set $\mathcal{V}$ of variables.

f) If φ is an $\mathrm{MSO}(\Sigma, S)$–sentence, then $\mathrm{Free}(\varphi) = \emptyset$ and thus $\llbracket \varphi \rrbracket \in S\langle\!\langle \Sigma^* \rangle\!\rangle$.

5.1.6 Remark. Each semiring S that is induced by a Boolean algebra has a natural complement operation $S \to S$, $x \mapsto \overline{x}$. Thus, for such a semiring we can simply set

$$\llbracket \neg\varphi \rrbracket_{\mathcal{V}}(w, \sigma) := \overline{\llbracket \varphi \rrbracket_{\mathcal{V}}(w, \sigma)}$$

for any $\mathrm{MSO}(\Sigma, S)$–formula φ and any $(w, \sigma) \in N_{\mathcal{V}}$, where $\mathcal{V}$ is a finite set of variables containing $\mathrm{Free}(\varphi)$. As a result, one obtains

$$\llbracket \varphi \wedge \psi \rrbracket_{\mathcal{V}} = \llbracket \neg(\neg\varphi \vee \neg\psi) \rrbracket_{\mathcal{V}},$$
$$\llbracket \forall x\, \varphi \rrbracket_{\mathcal{V}} = \llbracket \neg\exists x\, \varphi \rrbracket_{\mathcal{V}},$$
$$\llbracket \forall X \varphi \rrbracket_{\mathcal{V}} = \llbracket \neg\exists X \varphi \rrbracket_{\mathcal{V}}.$$

In particular, $\text{MSO}(\Sigma, S)$ can be viewed as multi-valued logic (cf. [6, page 73]). For the Boolean semiring $\mathbb{B}$, which is induced by the 2–valued Boolean algebra, the weighted logic $\text{MSO}(\Sigma, \mathbb{B})$ corresponds precisely to the classical logic $\text{MSO}(\Sigma)$ (see Proposition 5.1.10).

5.1.7 Remark. As we have indicated in Definition 5.1.4, we calculate the product

$$\llbracket \forall X \varphi \rrbracket (w, \sigma) = \prod_{I \subseteq \text{dom}(w)} \llbracket \varphi \rrbracket_{\mathcal{V} \cup \{X\}} (w, \sigma[X \mapsto I])$$

following some fixed enumeration of the power set $\mathcal{P}(\text{dom}(w))$. For instance, we may consider the enumeration induced by the lexicographic order on $\mathcal{P}(\text{dom}(w))$. More precisely, we consider the map

$$I \mapsto u_I \in \{0,1\}^{|w|} \text{ with } u_I(i) = \begin{cases} 1 & \text{if } i \in I \\ 0 & \text{otherwise} \end{cases} \quad (I \subseteq \text{dom}(w), i \in \text{dom}(w)),$$

which provides a bijective correspondence between the set $\mathcal{P}(\text{dom}(w))$ and the set of words over the alphabet $\{0,1\}$ of length $|w|$. We then define the lexicographic order $<_{\text{lex}}$ on $\{0,1\}^{|w|}$ naturally by

$$u <_{\text{lex}} u' \iff u(j) = 0 < 1 = u'(j) \text{ for } j = \min\{i \in \text{dom}(w) \mid u(i) \neq u'(i)\}$$

for any distinct words $u \neq u'$ in $\{0,1\}^{|w|}$. Via the bijection above, we can transfer this lexicographic order to the set $\mathcal{P}(\text{dom}(w))$ by setting

$$I <_{\text{lex}} J \iff u_I <_{\text{lex}} u_J$$

for any distinct subsets $I \neq J$ of $\text{dom}(w)$.

Observe that, given $\varphi \in \text{MSO}(\Sigma, S)$, we have defined semantics $\llbracket \varphi \rrbracket_{\mathcal{V}}$ for each finite set $\mathcal{V}$ of variables containing $\text{Free}(\varphi)$. The following result, which is proved inductively in [6, Proposition 3.3], shows that these semantics are consistent with each other.

5.1.8 Lemma. *Let φ be an $\text{MSO}(\Sigma, S)$–formula, $\mathcal{V} \subseteq \text{Var}$ a finite set of variables such that $\text{Free}(\varphi) \subseteq \mathcal{V}$ and $v = (w, \sigma) \in N_{\mathcal{V}}$. We have*

$$\llbracket \varphi \rrbracket_{\mathcal{V}}(w, \sigma) = \llbracket \varphi \rrbracket (w, \sigma|_{\text{Free}(\varphi)}).$$

In particular, this yields $\llbracket \varphi \rrbracket_{\mathcal{V}}(\varepsilon) = \llbracket \varphi \rrbracket(\varepsilon)$ for the empty word.

Hence, the coefficient $\llbracket \varphi \rrbracket_{\mathcal{V}}(w, \sigma)$ of the $\mathcal{V}$–semantics of $\varphi \in \text{MSO}(\Sigma, S)$ only depends on the word w and the restriction $\sigma|_{\text{Free}(\varphi)}$ of the $(\mathcal{V}, w)$–assignment σ to $\text{Free}(\varphi)$. In particular, Lemma 5.1.8 can be viewed as a weighted version of Lemma 2.4.10.

5.1.9 Definition. A series $r \in S\langle\!\langle \Sigma^* \rangle\!\rangle$ is called $\text{MSO}(\Sigma, S)$–**definable** if there exists an $\text{MSO}(\Sigma, S)$–sentence such that $r = \llbracket \varphi \rrbracket$. We denote the collection of all $\text{MSO}(\Sigma, S)$–definable series in $S\langle\!\langle \Sigma^* \rangle\!\rangle$ by $S^{\text{mso}}\langle\!\langle \Sigma^* \rangle\!\rangle$.

We have seen in the proof of Proposition 3.2.7 that weighted automata with weights in the Boolean semiring $\mathbb{B}$ correspond precisely to the classical automata. The proof of the next result shows that we have a similar correspondence between the weighted logic $\mathrm{MSO}(\Sigma, \mathbb{B})$ and the classical logic $\mathrm{MSO}(\Sigma)$.

5.1.10 Proposition. *The map*

$$\mathrm{char} \colon \mathcal{P}(\Sigma^*) \to \mathbb{B}\langle\!\langle \Sigma^* \rangle\!\rangle, \quad L \mapsto \mathbb{1}_L$$

and its inverse

$$\mathrm{supp} \colon \mathbb{B}\langle\!\langle \Sigma^* \rangle\!\rangle \to \mathcal{P}(\Sigma^*), \quad r \mapsto \mathrm{supp}(r)$$

provide a one-to-one correspondence between the languages that are definable by some $\mathrm{MSO}(\Sigma)$*–sentence and the set* $\mathbb{B}^{\mathrm{mso}}\langle\!\langle \Sigma^* \rangle\!\rangle$ *of power series that are the semantics of some* $\mathrm{MSO}(\Sigma, \mathbb{B})$*–sentence.*

Proof. We prove a more general correspondence for formulas and not just for sentences. Thus, we consider a finite set $\mathcal{V} \subseteq \mathrm{Var}$ of variables and formulas whose free variables are contained in $\mathcal{V}$. The statement of the proposition then corresponds to the special case $\mathcal{V} = \emptyset$.

First, we consider a classical $\mathrm{MSO}(\Sigma)$–formula φ with $\mathrm{Free}(\varphi) \subseteq \mathcal{V}$ and we show that the characteristic series $\mathbb{1}_{L_{\mathcal{V}}(\varphi)}$ coincides with the $\mathcal{V}$–semantics of some $\mathrm{MSO}(\Sigma, \mathbb{B})$–formula. By Remark 5.1.2b) we can assume without loss of generality that in φ negation is only applied to atomic formulas, i.e. we can understand φ also as weighted $\mathrm{MSO}(\Sigma, \mathbb{B})$–formula. Then one can show by structural induction that

$$[\![\varphi]\!]_{\mathcal{V}} = \mathbb{1}_{L_{\mathcal{V}}(\varphi)}.$$

A short justification of this identity for atomic formulas and their negations can be found in Remark 5.1.5b). For instance, if we consider first-order existential quantifications, then exploiting the arithmetic of the Boolean semiring $\mathbb{B}$ yields

$$[\![\exists x\, \varphi]\!]_{\mathcal{V}}(w, \sigma) = 1 \iff \sum_{i \in \mathrm{dom}(w)} [\![\varphi]\!]_{\mathcal{V} \cup \{x\}}(w, \sigma[x \mapsto i]) = 1$$

$$\iff \underbrace{[\![\varphi]\!]_{\mathcal{V} \cup \{x\}}(w, \sigma[x \mapsto i]) = 1}_{\substack{\iff (w, \sigma[x \mapsto i]) \in L_{\mathcal{V} \cup \{x\}}(\varphi) \\ \text{by induction hypothesis}}} \text{ for some } i \in \mathrm{dom}(w)$$

$$\iff (w, \sigma[x \mapsto i]) \models \varphi \text{ for some } i \in \mathrm{dom}(w)$$

$$\iff (w, \sigma) \models \exists x\, \varphi$$

$$\iff (w, \sigma) \in L_{\mathcal{V}}(\varphi)$$

for any $(w, \sigma) \in N_{\mathcal{V}}$, and $([\![\exists x\, \varphi]\!]_{\mathcal{V}}, v) = 0$ for any non-valid word $v \in \Sigma_{\mathcal{V}}^*$, i.e. for any $v \notin N_{\mathcal{V}}$. Similarly, one can verify the claim for disjunctions, conjunctions, first-order universal quantifications and the second-order quantifications. We indeed have to include conjunction and universal quantifications, since we want to restrict negations to atomic formulas.

Now let φ be an $\mathrm{MSO}(\Sigma, \mathbb{B})$–formula with $\mathrm{Free}(\varphi) \subseteq \mathcal{V}$. We claim that the support of its $\mathcal{V}$–semantics coincides with the language $L_\mathcal{V}(\psi)$ defined by some $\mathrm{MSO}(\Sigma)$–formula ψ, i.e. that there exists a $\psi \in \mathrm{MSO}(\Sigma)$ such that

$$L_\mathcal{V}(\psi) = \mathrm{supp}(\llbracket \varphi \rrbracket_\mathcal{V}).$$

Indeed, the weighted $\mathrm{MSO}(\Sigma, \mathbb{B})$–formulas 0 and 1 can be realized by the classical $\mathrm{MSO}(\Sigma)$–formulas $\bot$ and $\top$ or by the $\mathrm{MSO}(\Sigma)$–sentences $\exists x\, x < x$ and $\forall x\, x \le x$, respectively (see Remark A.2.6). More precisely, we have

$$L_\mathcal{V}(\bot) = L_\mathcal{V}(\exists x\, x < x) = \emptyset = \mathrm{supp}(\llbracket 0 \rrbracket_\mathcal{V}),$$
$$L_\mathcal{V}(\top) = L_\mathcal{V}(\forall x\, x \le x) = N_\mathcal{V} = \mathrm{supp}(\llbracket 1 \rrbracket_\mathcal{V}).$$

Thus, proceeding by structural induction, one can complete the proof. □

Hence, the structure of the Boolean semiring $\mathbb{B}$ perfectly models the qualitative manner of the classical monadic second-order logic for words over Σ. More precisely, the proof of Proposition 5.1.10 shows that we even have a correspondence between the classical $\mathrm{MSO}(\Sigma)$–formulas and the weighted $\mathrm{MSO}(\Sigma, \mathbb{B})$–formulas.

5.1.11 Definition. Given $\varphi \in \mathrm{MSO}(\Sigma, S)$, we define the set $\mathrm{const}(\varphi)$ of **constants** in the weighted formula φ by structural induction as follows:

$$\mathrm{const}(s) := \{s\},$$
$$\mathrm{const}(\varphi) := \emptyset \quad (\text{if } \varphi \text{ is atomic}),$$
$$\mathrm{const}(\neg\varphi) := \mathrm{const}(\varphi),$$
$$\mathrm{const}(\varphi \vee \psi) := \mathrm{const}(\varphi \wedge \psi) := (\mathrm{const}(\varphi) \cup \mathrm{const}(\psi)),$$
$$\mathrm{const}(\exists x\, \varphi) := \mathrm{const}(\forall x\, \varphi) := \mathrm{const}(\varphi),$$
$$\mathrm{const}(\exists X\varphi) := \mathrm{const}(\forall X\varphi) := \mathrm{const}(\varphi).$$

We note that $\mathrm{const}(\varphi)$ is a subset of the semiring S.

To illustrate the diversity of the weighted logic $\mathrm{MSO}(\Sigma, S)$, we consider once again the semirings from Example 3.1.5, which provide different interpretations of weights and thus different application scenarios for weighted $\mathrm{MSO}(\Sigma, S)$–formulas and their semantics (cf. Droste and Kuske [12, page 131], Droste and Gastin [10, page 183]).

5.1.12 Example.

a) The classical logic $\mathrm{MSO}(\Sigma)$ describes whether or not a certain property (e.g. "there exist two consecutive a's") holds for a given word. The weighted logic $\mathrm{MSO}(\Sigma, \mathbb{N})$ over the semiring $\mathbb{N}$ of natural numbers allows us to count "how often" this property holds. More precisely, let φ be an $\mathrm{MSO}(\Sigma, \mathbb{N})$–formula containing no constants or weights, i.e. $\mathrm{const}(\varphi) = \emptyset$. Then φ can be understood also as classical $\mathrm{MSO}(\Sigma)$–formula. Thus, for any $(w, \sigma) \in N_\mathcal{V}$ we can interpret the coefficient $\llbracket \varphi \rrbracket(w, \sigma)$ as the number of ways a machine could proceed to show that $(w, \sigma) \models \varphi$, i.e. as the number of proofs we have for $(w, \sigma) \models \varphi$. In fact, the machine could proceed inductively over the structure of φ:

- If φ is a literal, then the number of proofs for $(w,\sigma) \models \varphi$ is 0 or 1, depending on whether or not (w,σ) does model φ. Indeed, we have

$$[\![\varphi]\!](w,\sigma) = \begin{cases} 1 & \text{if } (w,\sigma) \models \varphi \\ 0 & \text{if } (w,\sigma) \not\models \varphi. \end{cases}$$

- If $[\![\varphi]\!](w,\sigma) = m$ and $[\![\psi]\!](w,\sigma) = n$, then the number of proofs for the claim $(w,\sigma) \models \varphi \vee \psi$ should be $m+n$, since any reason for $(w,\sigma) \models \varphi$ or $(w,\sigma) \models \psi$ suffices. On the other hand, to prove the claim $(w,\sigma) \models \varphi \wedge \psi$, the machine could arbitrarily pair the reasons for $(w,\sigma) \models \varphi$ and $(w,\sigma) \models \psi$, i.e. there are $m \cdot n$ possible proofs. In particular, if $m = 0$ or $n = 0$, then there is no way to show that $(w,\sigma) \models \varphi \wedge \psi$. Indeed, we have

$$[\![\varphi \vee \psi]\!] = m + n \text{ and } [\![\varphi \wedge \psi]\!] = m \cdot n.$$

- Similarly, the machine could deal with existential and universal quantifications.

For instance, we consider the sentence φ given by

$$\exists x \, \exists y \, \underbrace{(\mathrm{suc}(x,y) \wedge P_a(x) \wedge P_a(y))}_{=\psi}.$$

Then we have $w \models \varphi$ if and only if there exist two consecutive a's in the word w. For any word w, the coefficient $[\![\varphi]\!](w)$ evaluates to the sum

$$[\![\varphi]\!](w) = \sum_{i \in \mathrm{dom}(w)} \sum_{j \in \mathrm{dom}(w)} [\![\psi]\!](v_{i,j}),$$

where $v_{i,j}$ denotes the pair consisting of the word w and the $(\{x,y\},w)$–assignment σ with $\sigma(x) = i, \sigma(y) = j$, and the coefficient $[\![\psi]\!](v_{i,j})$ is determined by the product

$$[\![\psi]\!](v_{i,j}) = \underbrace{[\![\mathrm{suc}(x,y)]\!](v_{i,j})}_{\substack{=1 \\ \Leftrightarrow j=i+1}} \cdot \underbrace{[\![P_a(x)]\!](v_{i,j})}_{\substack{=1 \\ \Leftrightarrow w(i)=a}} \cdot \underbrace{[\![P_a(y)]\!](v_{i,j})}_{\substack{=1 \\ \Leftrightarrow w(j)=a}}$$

$$= \begin{cases} 1 & \text{if } j = i+1 \text{ and } w(i) = w(j) = a \\ 0 & \text{otherwise.} \end{cases}$$

Indeed, it requires a short computation to show that $[\![\mathrm{suc}(x,y)]\!] = \mathbb{1}_{L(\mathrm{suc}(x,y))}$. Thus, $[\![\varphi]\!](w)$ counts the number of pairs $(i,j) \in \mathrm{dom}(w)^2$ of positions such that

$$j = i+1 \text{ and } w(i) = w(j) = a,$$

each of which constitutes a proof for $w \models \varphi$. Considering e.g. the word $w = aaaa$, we obtain exactly three such pairs and thus $[\![\varphi]\!](aaaa) = 3$.

b) Weights coming from the tropical semiring $\text{Trop} = (\overline{\mathbb{R}}_{\geq 0}, \min, +, \infty, 0)$ can be viewed as resources or costs. Furthermore, we can regard the letters in Σ as actions and words as processes, i.e. as action sequences. Assuming that each action $a \in \Sigma$ is assigned a cost $c_a \in \overline{\mathbb{R}}_{\geq 0}$, we now consider the $\text{MSO}(\Sigma, \text{Trop})$–sentence

$$\varphi := \forall x \bigvee_{a \in \Sigma} (P_a(x) \wedge c_a).$$

For any $w \in \Sigma^*$ the coefficient $[\![\varphi]\!](w)$ then evaluates to

$$[\![\varphi]\!](w) = \sum_{i \in \text{dom}(w)} \min_{a \in \Sigma^*} \Big(\underbrace{[\![P_a(x)]\!](w, [x \mapsto i]) + [\![c_a]\!](w, [x \mapsto i])}_{= \begin{cases} 0 + c_a = c_a & \text{if } a = w(i) \\ \infty + c_a = \infty & \text{otherwise} \end{cases}} \Big)$$

$$= \sum_{i \in \text{dom}(w)} c_{w(i)}.$$

where $[x \mapsto i]$ denotes the $(\{x\}, w)$–assignment that maps x to i. Thus, the semantics $[\![\varphi]\!]$ assigns to each process the total costs of the actions it contains.

c) We interpret elements of the Viterbi semiring $\text{Prob} = ([0, 1], \max, \cdot, 0, 1)$ as probabilities and we assume that each letter $a \in \Sigma$ is assigned a reliability $r_a \in [0, 1]$. Moreover, we consider the $\text{MSO}(\Sigma, \text{Prob})$–sentence

$$\varphi := \forall x \bigvee_{a \in \Sigma} (P_a(x) \wedge r_a).$$

Then for any $w \in \Sigma^*$ the coefficient $[\![\varphi]\!](w)$ evaluates to

$$[\![\varphi]\!](w) = \prod_{i \in \text{dom}(w)} \max_{a \in \Sigma^*} \Big(\underbrace{[\![P_a(x)]\!](w, [x \mapsto i]) \cdot [\![r_a]\!](w, [x \mapsto i])}_{= \begin{cases} 1 \cdot r_a = r_a & \text{if } a = w(i) \\ 0 \cdot r_a = 0 & \text{otherwise} \end{cases}} \Big)$$

$$= \prod_{i \in \text{dom}(w)} r_{w(i)}.$$

where $[x \mapsto i]$ denotes the $(\{x\}, w)$–assignment that maps x to i. Thus, $[\![\varphi]\!](w)$ can be interpreted as the reliability of the word w.

d) For the Łukasiewicz semiring $S = ([0, 1], \max, \otimes, 0, 1)$, which occurs in the MV–algebra used to define the semantics of Łukasiewicz multi-valued logic, a restriction of Łukasiewicz logic coincides with the weighted logic $\text{MSO}(\Sigma, S)$ (cf. Droste and Gastin [10, page 183]).

Summarizing, weighted formulas can be used to determine quantitative properties, while classical formulas provide only qualitative statements. Moreover, the choice of the underlying semiring admits many different interpretations and applications of (the semantics of) weighted formulas.

130

In the subsequent example we examine further $\mathrm{MSO}(\Sigma, S)$–definable power series, with which we are already familiar, since we have encountered them in Example 3.2.11 and Example 3.3.11 when we showed their recognizability.

5.1.13 Example. Consider the alphabet $\Sigma = \{a, b\}$ and the $\mathrm{MSO}(\Sigma, S)$–sentence

$$\varphi := \exists x\, P_a(x).$$

If we regard φ as classical $\mathrm{MSO}(\Sigma)$–sentence, then it defines the language

$$L(\varphi) = \{w \in \Sigma^* \mid w \models \exists x P_a(x)\} = \{w \in \Sigma^* \mid |w|_a > 0\} = \Sigma^* \{a\} \Sigma^*,$$

which is recognizable. On the other hand, considered as weighted sentence, φ counts how often the letter a occurs in a word. More precisely, its semantics $[\![\varphi]\!] \in S\langle\!\langle \Sigma^* \rangle\!\rangle$ is given by

$$[\![\varphi]\!](w) = \sum_{i \in \mathrm{dom}(w)} [\![P_a(x)]\!](w, [x \mapsto i]) = \sum_{\substack{i \in \mathrm{dom}(w) \\ w(i) = a}} 1$$

for any $w \in \Sigma^*$, where $[x \mapsto i]$ denotes the $(\{x\}, w)$–assignment that maps x to i. This shows in particular that the term "how often" depends on the underlying semiring S:

- If $S = \mathbb{N}$ is the semiring of natural numbers, then the semantics of φ is given by the formal power series

$$[\![\varphi]\!] = \sum_{w \in \Sigma^*} \left(\sum_{\substack{i \in \mathrm{dom}(w) \\ w(i) = a}} 1 \right) w = \sum_{w \in \Sigma^*} |w|_a w,$$

  which is recognizable (see Example 3.2.11). Thus, the series $[\![\varphi]\!] = |\cdot|_a$ belongs to $\mathbb{N}^{\mathrm{mso}}\langle\!\langle \Sigma^* \rangle\!\rangle$ as well as to $\mathbb{N}^{\mathrm{rec}}\langle\!\langle \Sigma^* \rangle\!\rangle$.

- If $S = \mathbb{F}_2$ is the ring of integers modulo 2, then we compute

$$[\![\varphi]\!](w) = \sum_{\substack{i \in \mathrm{dom}(w) \\ w(i) = a}} 1 = \begin{cases} 0 & \text{if } |w|_a \text{ is even} \\ 1 & \text{if } |w|_a \text{ is odd} \end{cases}$$

  for any $w \in \Sigma^*$. We have seen in Example 3.3.11a) that $[\![\varphi]\!] \in \mathbb{F}_2^{\mathrm{rec}}\langle\!\langle \Sigma^* \rangle\!\rangle$.

- If $S = \mathbb{B}$ is the Boolean semiring, then for any $w \in \Sigma^*$ we obtain

$$[\![\varphi]\!] = \mathbb{1}_{L(\varphi)}$$

  by Proposition 5.1.10. We have seen in Example 3.3.11a) that $[\![\varphi]\!] \in \mathbb{B}^{\mathrm{rec}}\langle\!\langle \Sigma^* \rangle\!\rangle$.

We further observe that we have $\mathrm{supp}([\![\varphi]\!]) = \{w \in \Sigma^* \mid |w|_a \text{ is odd}\} \neq L(\varphi)$ for the semiring $\mathbb{F}_2$, while we obtain $\mathrm{supp}([\![\varphi]\!]) = \Sigma^* \{a\} \Sigma^* = L(\varphi)$ for the semirings $\mathbb{N}$ and $\mathbb{B}$.

Our goal in this chapter is to extend the Büchi–Elgot–Trakhtenbrot Theorem to the realm of formal power series, i.e. we want to show that a power series is recognizable if and only if it coincides with the semantics of some weighted sentence. Crucial for the proof of this characterization will be closure properties of $S^{\mathrm{rec}}\langle\!\langle\Sigma^*\rangle\!\rangle$ under the constructs of the weighted logic $\mathrm{MSO}(\Sigma, S)$. However, the following examples point out the need for restricting the weighted logic $\mathrm{MSO}(\Sigma, S)$. More precisely, they exhibit that in general neither conjunction nor universal quantifications preserve recognizability. The first example is due to Droste and Gastin [10, Example 3.4].

5.1.14 Example. Conjunction does not preserve recognizability, since $S^{\mathrm{rec}}\langle\!\langle\Sigma^*\rangle\!\rangle$ is in general not closed under the Hadamard product (see Remark 4.3.11). For instance, we consider the alphabet $\Sigma = \{a, b\}$, the semiring $S = (\mathcal{P}(\Sigma^*), \cup, \cdot, \emptyset, \{\varepsilon\})$ of languages over Σ and the $\mathrm{MSO}(\Sigma, S)$–sentence

$$\varphi := \forall x \, ((P_a(x) \wedge \{a\}) \vee (P_b(x) \wedge \{b\})).$$

Then for each $w \in \Sigma^*$ we obtain

$$[\![\varphi]\!](w) = \prod_{i \in \mathrm{dom}(w)} \{w(i)\} = \{w(1)\} \dots \{w(n)\} = \{w(1) \dots w(n)\} = \{w\}$$

where $n = |w|$. The linear representation (λ, μ, γ) of dimension $Q = \{1\}$ over Σ and S with $\lambda = \gamma = \{\varepsilon\}$ and $\mu(a) = \{a\}$ for each $a \in \Sigma$ fulfills

$$\begin{aligned}
||(\lambda, \mu, \gamma)||(w) &= \lambda \cdot \mu(w) \cdot \gamma \\
&= \{\varepsilon\} \cdot \{w(1)\} \dots \{w(n)\} \cdot \{\varepsilon\} \\
&= \{w(1) \dots w(n)\} = \{w\} = [\![\varphi]\!](w)
\end{aligned}$$

for any $w \in \Sigma^*$ with $n = |w|$. Thus, $[\![\varphi]\!]$ belongs to $S^{\mathrm{rec}}\langle\!\langle\Sigma^*\rangle\!\rangle$ by Proposition 3.3.8. On the other hand, one can show that the power series

$$[\![\varphi \wedge \varphi]\!] = [\![\varphi]\!] \odot [\![\varphi]\!]$$

is not recognizable (cf. [10, Example 3.4]). Hence, conjunction does in general not preserve recognizability.

5.1.15 Example. Based on [6, Example 3.4 and Example 3.5], we now examine universal quantifications in $\mathrm{MSO}(\Sigma, \mathbb{N})$.

a) If we consider the $\mathrm{MSO}(\Sigma, \mathbb{N})$–sentence $\forall x \, 2$, then for each $w \in \Sigma^*$ we obtain

$$[\![\forall x \, 2]\!](w) = \prod_{i \in \mathrm{dom}(w)} 2 = 2^{|w|}.$$

We claim that the semantics $[\![\forall x \, 2]\!] \in \mathbb{N}\langle\!\langle\Sigma^*\rangle\!\rangle$ coincides with the behavior of the weighted automaton $\mathcal{A}$ (with weights in $\mathbb{N}$) which is represented by the following state diagram:

$$1 \longrightarrow \boxed{q} \longrightarrow 1$$

with a self-loop labeled a / 2 on state q.

$$\text{(for any } a \in \Sigma\text{)}$$

Indeed, for any $w = a_1 \ldots a_n \in \Sigma^*$ the unique path in $\mathcal{A}$ with label w is given by $P_w = q a_1 q a_2 q \ldots a_n q$ and thus

$$
||\mathcal{A}||(w) = \sum_{\substack{P \in \text{Path}(\mathcal{A}) \\ \text{label}(P)=w}} \text{weight}(P) = \text{weight}(P_w)
$$

$$
= \underbrace{\text{in}(q)}_{=1} \cdot \prod_{i=1}^{n} \underbrace{\text{wt}(q, a_i, q)}_{=2} \cdot \underbrace{\text{out}(q)}_{=1}
$$

$$
= \prod_{i=1}^{n} 2 = 2^n = 2^{|w|}.
$$

Therefore, $[\![\forall x\, 2]\!]$ belongs to $\mathbb{N}^{\text{rec}}\langle\!\langle \Sigma^* \rangle\!\rangle$. However, we claim that the power series $[\![\forall y\, \forall x\, 2]\!]$ is not recognizable due to its growth. Suppose that $[\![\forall y\, \forall x\, 2]\!] = ||\mathcal{A}||$ for some weighted automaton $\mathcal{A} = (Q, \text{in}, \text{wt}, \text{out})$ and set

$$
M = \max\{|Q|, \text{in}(q), \text{wt}(p, a, q), \text{out}(q) \mid p, q \in Q, a \in \Sigma\}.
$$

Then for any $w \in \Sigma^*$ and any path P in $\mathcal{A}$ with label w we obtain

$$
\text{weight}(P) \leq M^{|w|+2}.
$$

Since there are exactly $|Q|^{|w|+1}$ paths labeled by w, this results in

$$
||\mathcal{A}||(w) = \sum_{\substack{P \in \text{Path}(\mathcal{A}) \\ \text{label}(P)=w}} \text{weight}(P)
$$

$$
\leq |Q|^{|w|+1} \cdot M^{|w|+2}
$$

$$
\leq M^{|w|+1} \cdot M^{|w|+2} = M^{2|w|+3},
$$

However, this contradicts

$$
||\mathcal{A}||(w) = [\![\forall y\, \forall x\, 2]\!](w) = \prod_{i \in \text{dom}(w)} \prod_{j \in \text{dom}(w)} 2 = (2^{|w|})^{|w|} = 2^{|w|^2},
$$

for any word w fulfilling $|w|^2 > \log_2(M) \cdot (2|w| + 3)$, since in this case we have

$$
2^{|w|^2} > 2^{\log_2(M) \cdot (2|w|+3)} = \left(2^{\log_2(M)}\right)^{2|w|+3} = M^{2|w|+3}.
$$

This shows that first-order universal quantification does in general not preserve recognizability.

$\cdot$

b) The semantics of the weighted $MSO(\Sigma, \mathbb{N})$–sentence 2 is given by the constant power series

$$[\![2]\!] = \sum_{w \in \Sigma^*} 2w,$$

which is recognizable (see Remark 4.3.12) and thus belongs to $\mathbb{N}^{rec}\langle\!\langle \Sigma^* \rangle\!\rangle$. However, as above one can show that the power series $[\![\forall X 2]\!]$, with

$$[\![\forall X 2]\!](w) = \prod_{I \subseteq \mathrm{dom}(w)} 2 = 2^{2^{|w|}}$$

for any $w \in \Sigma^*$, is not recognizable due to its growth. Hence, second-order universal quantification does in general not preserve recognizability.

We observe that the underlying semiring S in Example 5.1.14 is not commutative. However, $S^{rec}\langle\!\langle \Sigma^* \rangle\!\rangle$ is closed under the Hadamard product and thus under conjunction if the semiring S is commutative (see Proposition 4.3.10 and Remark 4.3.11). Therefore, we can bypass the difficulties with conjunction by considering only commutative semirings. Example 5.1.15 shows that, even for commutative semirings like the natural numbers, unrestricted universal quantifications are too strong to preserve recognizability. This implies that the weighted logic $MSO(\Sigma, S)$ is more expressive than weighted automata. Therefore, we have to work with fragments or restrictions of $MSO(\Sigma, S)$ in order to obtain a further characterization of recognizability in the context of formal power series. To define the fragment of $MSO(\Sigma, S)$ we will consider in the following, we first have to introduce the notion of recognizable step functions.

5.1.16 Definition. A power series $r \in S\langle\!\langle \Sigma^* \rangle\!\rangle$ is called **recognizable step function** if there exists an $n \in \mathbb{N}$, scalars $s_1, \ldots, s_n \in S$ and recognizable languages $L_1, \ldots, L_n$ over Σ such that

$$r = \sum_{i=1}^{n} s_i \cdot \mathbb{1}_{L_i}.$$

The notion of a *recognizable* step function is reasonable, since by Proposition 3.2.8 the characteristic series of any recognizable language is recognizable, and by Proposition 4.3.3 and Remark 4.3.12 addition and scalar multiplication preserve recognizability. Hence, each recognizable step function is indeed a recognizable power series.

5.1.17 Remark. Given a recognizable step function

$$r = \sum_{i=1}^{n} s_i \cdot \mathbb{1}_{L_i}$$

the languages $L_1, \ldots, L_n$ can be assumed to form a partition of Σ^*, since the language theoretic operations union, intersection and (relative) complement preserve recognizability (see Proposition 2.3.4). For instance, we consider $n = 2$ and the recognizable step function

$$r = s_1 \cdot \mathbb{1}_{L_1} + s_2 \cdot \mathbb{1}_{L_2}.$$

Then r can be written as

$$r = 0 \cdot \mathbb{1}_{L_0} + s_1 \cdot \mathbb{1}_{L_1'} + s_2 \cdot \mathbb{1}_{L_2'} + (s_1 + s_2) \cdot \mathbb{1}_{L_{12}}$$

with $L_0 = \Sigma^* \setminus (L_1 \cup L_2), L_1' = L_1 \setminus L_2, L_2' = L_2 \setminus L_1$ and $L_{12} = L_1 \cap L_2$. The languages L_0, L_1', L_2', L_{12} form a partition of Σ^*. In particular, we infer from this that a recognizable step function takes only finitely many values in S.

5.1.18 Definition. An $\mathrm{MSO}(\Sigma, S)$–formula φ is called **restricted** if it contains no second-order universal quantification and whenever φ contains a first-order universal quantification of the form $\forall x\, \psi$, then $[\![\psi]\!]$ is a recognizable step function. More precisely, we define the set of restricted $\mathrm{MSO}(\Sigma, S)$–formulas inductively as follows:

(1) The constant s is a restricted $\mathrm{MSO}(\Sigma, S)$–formula for any $s \in S$.

(2) All literals in $\mathrm{MSO}(\Sigma, S)$ are restricted.

(3) If φ and ψ are restricted $\mathrm{MSO}(\Sigma, S)$–formulas, then so are their disjunction $(\varphi \vee \psi)$ and conjunction $(\varphi \wedge \psi)$.

(4) If φ is a restricted $\mathrm{MSO}(\Sigma, S)$–formula, $x \in \mathrm{Var}_1$ and $X \in \mathrm{Var}_2$, then the existential quantifications $\exists x\, \varphi$ and $\exists X \varphi$ are again restricted.

(5) If φ is a restricted $\mathrm{MSO}(\Sigma, S)$–formula such that its semantics $[\![\varphi]\!]$ is a recognizable step function, then also the first-order universal quantification $\forall x\, \varphi$ is restricted.

We denote by $\mathrm{RMSO}(\Sigma, S)$ the collection of all restricted $\mathrm{MSO}(\Sigma, S)$–formulas.

We observe that due to (5) this is not a purely syntactic definition, since the restriction of first-order universal quantification is on the semantics of formulas, but one can show that under certain suitable assumptions on the semiring it is decidable whether or not a given $\mathrm{MSO}(\Sigma, S)$–formula is restricted (cf. [6, Proposition 4.6]). On the other hand, we want to mention that Droste and Gastin developed also a purely syntactical approach. More precisely, in [10] they consider the fragment of *syntactically restricted* $\mathrm{MSO}(\Sigma, S)$–formulas. However, in this work we follow their first approach, which is presented in [5] and [6].

5.1.19 Definition. Let φ be a weighted $\mathrm{MSO}(\Sigma, S)$–formula. We call φ **existential** if it is of the form $\exists X_0, \ldots, X_n \psi$ for some $n \in \mathbb{N}$ and an $\mathrm{MSO}(\Sigma, S)$–formula ψ containing no second-order quantifications. We denote the collection of all existential $\mathrm{MSO}(\Sigma, S)$–formulas by $\mathrm{EMSO}(\Sigma, S)$ and the collection of all restricted existential $\mathrm{MSO}(\Sigma, S)$–formulas by $\mathrm{REMSO}(\Sigma, S)$.

5.1.20 Definition. Let $F \subseteq \mathrm{MSO}(\Sigma, S)$ be some fragment of $\mathrm{MSO}(\Sigma, S)$. A formal power series $r \in S\langle\!\langle \Sigma^* \rangle\!\rangle$ is called F–**definable** if there exists a weighted $\mathrm{MSO}(\Sigma, S)$–sentence $\varphi \in F$ such that $r = [\![\varphi]\!]$.

5.1.21 Notation. We set

$$S^{\mathrm{mso}}\langle\!\langle\Sigma^*\rangle\!\rangle := \{r \in S\langle\!\langle\Sigma^*\rangle\!\rangle \mid r \text{ is MSO}(\Sigma, S)\text{--definable}\},$$
$$S^{\mathrm{rmso}}\langle\!\langle\Sigma^*\rangle\!\rangle := \{r \in S\langle\!\langle\Sigma^*\rangle\!\rangle \mid r \text{ is RMSO}(\Sigma, S)\text{--definable}\},$$
$$S^{\mathrm{emso}}\langle\!\langle\Sigma^*\rangle\!\rangle := \{r \in S\langle\!\langle\Sigma^*\rangle\!\rangle \mid r \text{ is EMSO}(\Sigma, S)\text{--definable}\},$$
$$S^{\mathrm{remso}}\langle\!\langle\Sigma^*\rangle\!\rangle := \{r \in S\langle\!\langle\Sigma^*\rangle\!\rangle \mid r \text{ is REMSO}(\Sigma, S)\text{--definable}\}.$$

5.1.22 Remark. We consider again the Boolean semiring $\mathbb{B}$. By Proposition 5.1.10 we have

$$\mathbb{B}^{\mathrm{mso}}\langle\!\langle\Sigma^*\rangle\!\rangle = \{\mathbb{1}_L \mid L \subseteq \Sigma^* \text{ is definable by some MSO}(\Sigma)\text{--sentence}\}.$$

Similarly, we obtain

$$\mathbb{B}^{\mathrm{emso}}\langle\!\langle\Sigma^*\rangle\!\rangle = \{\mathbb{1}_L \mid L \subseteq \Sigma^* \text{ is definable by some existential MSO}(\Sigma)\text{--sentence}\}.$$

Furthermore, Proposition 3.2.7 implies that

$$\mathbb{B}^{\mathrm{rec}}\langle\!\langle\Sigma^*\rangle\!\rangle = \{\mathbb{1}_L \mid L \subseteq \Sigma^* \text{ is recognized by some classical automaton}\}.$$

Therefore, we achieve

$$\mathbb{B}^{\mathrm{rec}}\langle\!\langle\Sigma^*\rangle\!\rangle = \mathbb{B}^{\mathrm{mso}}\langle\!\langle\Sigma^*\rangle\!\rangle = \mathbb{B}^{\mathrm{emso}}\langle\!\langle\Sigma^*\rangle\!\rangle$$

by Theorem 2.5.1 and Corollary 2.5.14.

The aim of the next section is to prove that for an arbitrary commutative semiring the fragments RMSO(Σ, S) and REMSO(Σ, S) have the same expressive power as weighted automata, i.e. to establish the equalities

$$S^{\mathrm{rec}}\langle\!\langle\Sigma^*\rangle\!\rangle = S^{\mathrm{rmso}}\langle\!\langle\Sigma^*\rangle\!\rangle = S^{\mathrm{remso}}\langle\!\langle\Sigma^*\rangle\!\rangle.$$

5.2. Results of Droste and Gastin

In this section, we carry out the main steps of Droste and Gastin [6, § 4, § 5 and § 6] in order to establish a further characterization of recognizable power series by means of their weighted monadic second-order logic. As we have seen in the previous section, for non-commutative semirings conjunction does in general not preserve recognizability. Therefore, we restrict ourselves to commutative semirings in the following. Our aim is to demonstrate that, for commutative semirings, the behaviors of weighted automata are precisely the formal power series which are definable by particular weighted sentences. To this end, we first prove that any restricted weighted formula has as semantics a recognizable power series. Subsequently, we show that the behavior of any weighted automaton can be defined by a restricted existential weighted sentence. Ultimately, we consider the class of *locally finite* semirings. For this large class of semirings, we derive that the semantics of all weighted formulas – not only of the restricted ones – are recognizable power series. Hence, we establish quantitative extensions of the Büchi–Elgot–Trakhtenbrot Theorem.

5.2.1 Theorem (Droste and Gastin). *Let S be commutative and let $r \in S\langle\!\langle \Sigma^* \rangle\!\rangle$ be a formal power series. Then the following conditions are equivalent:*

(i) The power series r is recognizable.

(ii) The power series r is RMSO(Σ, S)–definable.

(iii) The power series r is REMSO(Σ, S)–definable.

This means $S^{\mathrm{rec}}\langle\!\langle \Sigma^* \rangle\!\rangle = S^{\mathrm{rmso}}\langle\!\langle \Sigma^* \rangle\!\rangle = S^{\mathrm{remso}}\langle\!\langle \Sigma^* \rangle\!\rangle$.

For their purely syntactical approach, Droste and Gastin established a similar result, which in contrast to Theorem 5.2.1 applies to arbitrary semirings, i.e. even to non-commutative ones (cf. [10, Theorem 4.7]).

The following result will help us in the proof of Theorem 5.2.1 and can be considered as an extension of Lemma 2.5.7 to the weighted setting. For its proof, we follow [6, Proposition 3.3].

5.2.2 Lemma. *Let φ be a weighted MSO(Σ, S)–formula and let $\mathcal{V} \subseteq \mathrm{Var}$ a finite set of variables such that Free$(\varphi) \subseteq \mathcal{V}$. Then $[\![\varphi]\!]_{\mathcal{V}}$ is recognizable if and only if $[\![\varphi]\!]$ is recognizable. Moreover, $[\![\varphi]\!]_{\mathcal{V}}$ is a recognizable step function if and only if $[\![\varphi]\!]$ is a recognizable step function.*

Proof. We let $\pi \colon \Sigma_{\mathcal{V}}^* \to \Sigma_{\varphi}^*$ be the unique length-preserving monoid homomorphism extending the map

$$\Sigma_{\mathcal{V}} \to \Sigma_{\varphi}, \ (a, \beta) \mapsto (a, \beta|_{\mathrm{Free}(\varphi)})$$

(see Lemma 2.1.12 and Remark 2.1.15). Then it is straightforward to verify that

$$\pi(w, \sigma) = (w, \sigma|_{\mathrm{Free}(\varphi)})$$

for any $(w, \sigma) \in N_{\mathcal{V}}$, i.e. π can be considered as a projection. If we use the transformation of power series induced by π (see Definition 3.2.13a)) and apply Lemma 5.1.8, then we obtain

$$[\![\varphi]\!]_{\mathcal{V}}(w, \sigma) = [\![\varphi]\!](w, \sigma|_{\mathrm{Free}(\varphi)})$$
$$= [\![\varphi]\!](\pi(w, \sigma))$$
$$= \left(\pi^{-1}\left([\![\varphi]\!]\right)\right)(w, \sigma)$$

for any $(w, \sigma) \in N_{\mathcal{V}}$. On the other hand, by definition we have $[\![\varphi]\!]_{\mathcal{V}}(v) = 0$ for any non-valid word $v \in \Sigma_{\mathcal{V}}^*$, i.e. for any $v \notin N_{\mathcal{V}}$. Summarizing, this results in

$$[\![\varphi]\!]_{\mathcal{V}} = \pi^{-1}([\![\varphi]\!]) \odot \mathbb{1}_{N_{\mathcal{V}}}.$$

We recall that the language $N_{\mathcal{V}}$ is recognizable (see Lemma 2.5.8). Hence, if $[\![\varphi]\!]$ is recognizable, then also the Hadamard product $[\![\varphi]\!]_{\mathcal{V}} = \pi^{-1}([\![\varphi]\!]) \odot \mathbb{1}_{N_{\mathcal{V}}}$ is recognizable by Proposition 3.2.15, Proposition 4.3.10 and Remark 4.3.11.

For the converse, we consider the language

$$F = \{\varepsilon\} \cup \{\varepsilon \neq (w, \sigma) \in N_\mathcal{V} \mid \sigma(x) = 1 \text{ for any } x \in \mathcal{V}_1 \setminus \text{Free}(\varphi),$$
$$\sigma(X) = \{1\} \text{ for any } X \in \mathcal{V}_2 \setminus \text{Free}(\varphi)\}$$

over the extended alphabet $\Sigma_\mathcal{V}$. Similar to the proof of Lemma 2.5.8, one can show that F is a recognizable language. Furthermore, for each $(w, \sigma) \in N_\varphi$ there exists a unique element $(w, \hat{\sigma}) \in F$ such that

$$\pi(w, \hat{\sigma}) = (w, \hat{\sigma}\big|_{\text{Free}(\varphi)}) = (w, \sigma).$$

Hence, using the transformation of power series induced by π (see Definition 3.2.13b)) and applying Lemma 5.1.8, we obtain

$$(\pi(\llbracket\varphi\rrbracket_\mathcal{V} \odot \mathbf{1}_F))(w, \sigma) = \sum_{u \in \pi^{-1}(w,\sigma)} \underbrace{(\llbracket\varphi\rrbracket_\mathcal{V} \odot \mathbf{1}_F)(u)}_{=0 \text{ if } u \notin F}$$
$$= \llbracket\varphi\rrbracket_\mathcal{V}(w, \hat{\sigma})$$
$$= \llbracket\varphi\rrbracket(w, \hat{\sigma}\big|_{\text{Free}(\varphi)})$$
$$= \llbracket\varphi\rrbracket(w, \sigma)$$

for any $(w, \sigma) \in N_\varphi$. On the other hand, given a non-valid word $v \in \Sigma_\varphi^*$, i.e. $v \notin N_\varphi$, any preimage $u \in \pi^{-1}(v) \subseteq \Sigma_\mathcal{V}^*$ of v under the projection π is again non-valid and does in particular not belong to the language F. This implies

$$(\pi(\llbracket\varphi\rrbracket_\mathcal{V} \odot \mathbf{1}_F))(v) = \sum_{u \in \pi^{-1}(v)} \underbrace{(\llbracket\varphi\rrbracket_\mathcal{V} \odot \mathbf{1}_F)(u)}_{=0 \text{ since } u \notin F}$$
$$= 0$$
$$= \llbracket\varphi\rrbracket(v).$$

Hence, if $\llbracket\varphi\rrbracket_\mathcal{V}$ is recognizable, then also

$$\llbracket\varphi\rrbracket = \pi(\llbracket\varphi\rrbracket_\mathcal{V} \odot \mathbf{1}_F)$$

is a recognizable power series by Proposition 3.2.8, Proposition 4.3.10, Remark 4.3.11 and Proposition 3.2.16.

A proof of the second part of the claim regarding recognizable step functions can be found in Droste and Gastin [10, Proposition 3.3]. $\qquad\square$

Recall that in the proof of the classical result of Büchi, Elgot and Trakhtenbrot we proceeded as follows: First, given a classical automaton $\mathcal{A}$, we directly constructed an explicit MSO(Σ)–sentence φ such that $L(\varphi) = L(\mathcal{A})$ (see Proposition 2.5.2). For the other implication, we proceeded by structural induction and showed the stronger fact that $L_\mathcal{V}(\varphi)$ is recognizable for any MSO(Σ)–formula with Free(φ) $\subseteq \mathcal{V}$ (see Proposition 2.5.12). Now we proceed similarly to prove Theorem 5.2.1. We start by proving the inclusion

$$S^{\text{rmso}}\langle\langle\Sigma^*\rangle\rangle \subseteq S^{\text{rec}}\langle\langle\Sigma^*\rangle\rangle.$$

From Weighted Formulas to Weighted Automata

Following [6, § 4], we proceed by structural induction to show the stronger fact that the $\mathcal{V}$–semantics $[\![\varphi]\!]_\mathcal{V}$ is recognizable for any $\mathrm{MSO}(\Sigma, S)$–formula φ and any finite set $\mathcal{V}$ of variables containing $\mathrm{Free}(\varphi)$.

5.2.3 Proposition. *Let $\varphi \in \mathrm{MSO}(\Sigma, S)$ be a literal or of the form $\varphi = s$ for an $s \in S$. Then its semantics $[\![\varphi]\!]$ is recognizable.*

Proof. If $\varphi = s$ for an $s \in S$, then its semantics is given by the constant power series

$$[\![\varphi]\!] = [\![s]\!] = \sum_{w \in \Sigma^*} sw = s \cdot \mathbb{1}_{\Sigma^*},$$

and therefore recognizable (see Remark 4.3.12). If φ is a literal, i.e. an atomic formula or the negation of an atomic formula, then we can understand φ also as classical $\mathrm{MSO}(\Sigma)$–formula to obtain

$$[\![\varphi]\!] = \mathbb{1}_{L(\varphi)} \in S\langle\!\langle \Sigma_\varphi^* \rangle\!\rangle$$

(see Remark 5.1.5b)). As the language $L(\varphi)$ is recognizable by Proposition 2.5.12, the power series $[\![\varphi]\!] = \mathbb{1}_{L(\varphi)}$ is recognizable by Proposition 3.2.8. □

5.2.4 Proposition. *Let φ and ψ be $\mathrm{MSO}(\Sigma, S)$–formulas such that $[\![\varphi]\!]$ and $[\![\psi]\!]$ are recognizable. Then the series $[\![\varphi \vee \psi]\!]$ is again recognizable. If the semiring S is commutative, then also the series $[\![\varphi \wedge \psi]\!]$ is recognizable.*

Proof. For the set $\mathcal{V} = \mathrm{Free}(\varphi) \cup \mathrm{Free}(\psi) = \mathrm{Free}(\varphi \vee \psi) = \mathrm{Free}(\varphi \wedge \psi)$, we have

$$[\![\varphi \vee \psi]\!] = [\![\varphi]\!]_\mathcal{V} + [\![\psi]\!]_\mathcal{V},$$
$$[\![\varphi \wedge \psi]\!] = [\![\varphi]\!]_\mathcal{V} \odot [\![\psi]\!]_\mathcal{V}.$$

As the power series $[\![\varphi]\!]_\mathcal{V}$ and $[\![\psi]\!]_\mathcal{V}$ are recognizable by Lemma 5.2.2, the claim follows by Proposition 4.3.3 and Proposition 4.3.10. □

As we apply Proposition 4.3.10 in the proof of Proposition 5.2.4 to show that the semantics $[\![\varphi \wedge \psi]\!] = [\![\varphi]\!]_\mathcal{V} \odot [\![\psi]\!]_\mathcal{V}$ is recognizable, it suffices to assume that

$$[\![\varphi]\!]_\mathcal{V}(w, \sigma) \cdot [\![\psi]\!]_\mathcal{V}(v, \tau) = [\![\psi]\!]_\mathcal{V}(v, \tau) \cdot [\![\varphi]\!]_\mathcal{V}(w, \sigma)$$

for any $(w, \sigma), (v, \tau) \in N_\mathcal{V}$, instead of supposing that the whole semiring S is commutative. One can even show that it suffices to assume that the sets $\mathrm{const}(\varphi)$ and $\mathrm{const}(\psi)$ of constants commute elementwise (cf. Droste and Gastin [10, Lemma 5.3]).

5.2.5 Proposition. *Let φ be an $\mathrm{MSO}(\Sigma, S)$–formula such that its semantics $[\![\varphi]\!]$ is recognizable. Then the series $[\![\exists x\, \varphi]\!]$ and $[\![\exists X \varphi]\!]$ are recognizable as well.*

Proof. First, we consider the second-order existential quantification $\exists X\varphi$. As in the proof of Lemma 2.5.11, we put $\mathcal{V} = \mathrm{Free}(\exists X\varphi)$, i.e. $X \notin \mathcal{V}$, and we consider the unique length-preserving monoid homomorphism $\pi \colon \Sigma_{\mathcal{V} \cup \{X\}}^* \to \Sigma_\mathcal{V}^*$ extending the map

$$\Sigma_{\mathcal{V} \cup \{X\}} \to \Sigma_\mathcal{V}, \ (a, \beta) \mapsto (a, \beta|_\mathcal{V})$$

(see Lemma 2.1.12 and Remark 2.1.15). As usual, we may view π as a projection.

Given $(w, \sigma) \in N_{\mathcal{V}}$, it is straightforward to verify that

$$\pi^{-1}(w, \sigma) = \{(w, \sigma[X \mapsto I]) \mid I \subseteq \text{dom}(w)\} \subseteq N_{\mathcal{V} \cup \{X\}}.$$

As a consequence, we obtain

$$\left(\pi\left([\![\varphi]\!]_{\mathcal{V} \cup \{X\}}\right)\right)(w, \sigma) = \sum_{u \in \pi^{-1}(w, \sigma)} [\![\varphi]\!]_{\mathcal{V} \cup \{X\}}(u)$$

$$= \sum_{I \subseteq \text{dom}(w)} [\![\varphi]\!]_{\mathcal{V} \cup \{X\}}(w, \sigma[X \mapsto I])$$

$$= [\![\exists X \varphi]\!](w, \sigma),$$

using the transformation of formal power series induced by π (see Definition 3.2.13b)). In particular, we have $\pi^{-1}(\varepsilon) = \{\varepsilon\}$ and

$$\left(\pi\left([\![\varphi]\!]_{\mathcal{V} \cup \{X\}}\right)\right)(\varepsilon) = [\![\varphi]\!]_{\mathcal{V} \cup \{X\}}(\varepsilon) = [\![\exists X \varphi]\!](\varepsilon)$$

for the empty word, which also belongs to $N_{\mathcal{V}}$ (see Remark 5.1.5e)). On the other hand, given a non-valid word $v \in \Sigma_{\mathcal{V}}^*$, i.e. $v \notin N_{\mathcal{V}}$, any preimage $u \in \pi^{-1}(v)$ of v under the projection π is again non-valid, yielding

$$\left(\pi\left([\![\varphi]\!]_{\mathcal{V} \cup \{X\}}\right)\right)(v) = \sum_{u \in \pi^{-1}(v)} \underbrace{[\![\varphi]\!]_{\mathcal{V} \cup \{X\}}(u)}_{\substack{=0 \\ \text{since } u \notin N_{\mathcal{V} \cup \{X\}}}}$$

$$= 0$$

$$= [\![\exists X \varphi]\!](v)$$

by definition of the weighted semantics. Summarizing, we have

$$[\![\exists X \varphi]\!] = \pi\left([\![\varphi]\!]_{\mathcal{V} \cup \{X\}}\right).$$

As $\text{Free}(\varphi) \subseteq \mathcal{V} \cup \{X\}$, the series $[\![\varphi]\!]_{\mathcal{V} \cup \{X\}}$ is recognizable by Lemma 5.2.2. Hence, the series $[\![\exists X \varphi]\!]$ is recognizable by Proposition 3.2.16.

Next, we turn to the first-order existential quantification $\exists x \, \varphi$. Similar to above, we now put $\mathcal{V} = \text{Free}(\exists x \, \varphi)$, i.e. $x \notin \mathcal{V}$, and we consider the unique length-preserving monoid homomorphism $\pi \colon \Sigma_{\mathcal{V} \cup \{x\}}^* \to \Sigma_{\mathcal{V}}^*$ extending the map

$$\Sigma_{\mathcal{V} \cup \{x\}} \to \Sigma_{\mathcal{V}}, \ (a, \beta) \mapsto (a, \beta|_{\mathcal{V}})$$

(see Lemma 2.1.12 and Remark 2.1.15). Given $\varepsilon \neq (w, \sigma) \in N_{\mathcal{V}}$, it is straightforward to verify that

$$\pi^{-1}(w, \sigma) \cap N_{\mathcal{V} \cup \{x\}} = \{(w, \sigma[x \mapsto i]) \mid i \in \text{dom}(w)\}.$$

Hence, we obtain

$$
\begin{aligned}
\left(\pi\left(\llbracket\varphi\rrbracket_{\mathcal{V}\cup\{x\}}\right)\right)(w,\sigma) &= \sum_{u\in\pi^{-1}(w,\sigma)} \underbrace{\llbracket\varphi\rrbracket_{\mathcal{V}\cup\{x\}}(u)}_{\substack{=0 \\ \text{if } u\notin N_{\mathcal{V}\cup\{x\}}}} \\
&= \sum_{i\in\mathrm{dom}(w)} \llbracket\varphi\rrbracket_{\mathcal{V}\cup\{x\}}(w,\sigma[x\mapsto i]) \\
&= \llbracket\exists x\,\varphi\rrbracket(w,\sigma),
\end{aligned}
$$

using the transformation of formal power series induced by π (see Definition 3.2.13b)). On the other hand, one can show as above that

$$
\left(\pi\left(\llbracket\varphi\rrbracket_{\mathcal{V}\cup\{x\}}\right)\right)(v) = \llbracket\exists x\,\varphi\rrbracket(v)
$$

for any non-valid word $v\in\Sigma_{\mathcal{V}}^*$, i.e. for any $v\notin N_{\mathcal{V}}$. Therefore, we obtain

$$
\llbracket\exists x\,\varphi\rrbracket = \pi\left(\llbracket\varphi\rrbracket_{\mathcal{V}\cup\{x\}}\right)\odot\mathbb{1}_{\Sigma_{\mathcal{V}}^+},
$$

since $\llbracket\exists x\,\varphi\rrbracket(\varepsilon)=0$ (see Remark 5.1.5e)). As above, one can prove that $\pi\left(\llbracket\varphi\rrbracket_{\mathcal{V}\cup\{x\}}\right)$ is a recognizable series. The characteristic series $\mathbb{1}_{\Sigma_{\mathcal{V}}^+}$ is recognizable by Proposition 3.2.8, since the language $\Sigma_{\mathcal{V}}^+ = \Sigma_{\mathcal{V}}^*\setminus\{\varepsilon\}$ is recognizable by Proposition 2.3.4. Hence, the series $\llbracket\exists x\,\varphi\rrbracket$ is recognizable by Proposition 4.3.10 and Remark 4.3.11, completing the proof. $\square$

5.2.6 Remark. In the proof of [6, Lemma 4.3] one finds the equality

$$
\llbracket\exists x\,\varphi\rrbracket = \pi\left(\llbracket\varphi\rrbracket_{\mathcal{V}\cup\{x\}}\right).
$$

However, in the proof of Proposition 5.2.5 we have restricted the series $\pi\left(\llbracket\varphi\rrbracket_{\mathcal{V}\cup\{x\}}\right)$ to the recognizable language $\Sigma_{\mathcal{V}}^+$ to obtain

$$
\llbracket\exists x\,\varphi\rrbracket = \pi\left(\llbracket\varphi\rrbracket_{\mathcal{V}\cup\{x\}}\right)\odot\mathbb{1}_{\Sigma_{\mathcal{V}}^+}.
$$

In fact, this restriction is necessary, as in general we have

$$
\begin{aligned}
\left(\pi\left(\llbracket\varphi\rrbracket_{\mathcal{V}\cup\{x\}}\right)\right)(\varepsilon) &= \sum_{u\in\pi^{-1}(\varepsilon)} \llbracket\varphi\rrbracket_{\mathcal{V}\cup\{x\}}(u) \\
&= \llbracket\varphi\rrbracket_{\mathcal{V}\cup\{x\}}(\varepsilon) \\
&\neq 0
\end{aligned}
$$

whereas $\llbracket\exists x\,\varphi\rrbracket(\varepsilon)=0$.

Unlike in the unweighted setting, in general we cannot express weighted universal quantification by existential quantification. Therefore, the next result, concerning first-order universal quantification, requires a crucial new construction of weighted automata. We present the proof of Droste and Gastin [10, Lemma 5.4], which uses our generalized statement of the Büchi–Elgot–Trakhtenbrot Theorem (see Theorem 2.5.19). An alternative proof, which however requires that the underlying semiring S is commutative, can be found in [6, Lemma 4.4].

5.2.7 Proposition. *Let φ be an $\mathrm{MSO}(\Sigma, S)$–formula such that its semantics $[\![\varphi]\!]$ is a recognizable step function. Then the series $[\![\forall x\, \varphi]\!]$ is again recognizable.*

Proof. We set $\mathcal{W} = \mathrm{Free}(\varphi) \cup \{x\}$ and $\mathcal{V} = \mathrm{Free}(\forall x\, \varphi) = \mathcal{W} \setminus \{x\}$. As $\mathrm{Free}(\varphi) \subseteq \mathcal{W}$, the series $[\![\varphi]\!]_{\mathcal{W}}$ is a recognizable step function by Lemma 5.2.2. Thus, we may write

$$[\![\varphi]\!]_{\mathcal{W}} = \sum_{j=1}^{n} s_j \cdot \mathbf{1}_{L_j},$$

where $n \in \mathbb{N}$, $s_1, \ldots, s_n$, and $L_1, \ldots, L_n$ are recognizable languages over the extended alphabet $\Sigma_{\mathcal{W}}$. Without loss of generality, we may assume that the languages $L_1, \ldots, L_n$ form a partition of $\Sigma_{\mathcal{V}}^*$ (see Remark 5.1.17). Moreover, we recall that $[\![\varphi]\!]_{\mathcal{W}}(v) = 0$ for any non-valid $v \in \Sigma_{\mathcal{W}}^*$, i.e. for any $v \notin N_{\mathcal{W}}$.

Now we consider the alphabet $\widetilde{\Sigma} = \{1, \ldots, n\} \times \Sigma$. Then words over the extended alphabet $\widetilde{\Sigma}_{\mathcal{V}} = \{1, \ldots, n\} \times \Sigma \times \{0, 1\}^{\mathcal{V}}$ are of the form

$$(\kappa, v) = (j_1, a_1, \beta_1) \ldots (j_m, a_m, \beta_m),$$

where $m \in \mathbb{N}$, $v = (a_1, \beta_1) \ldots (a_m, \beta_m) \in \Sigma_{\mathcal{V}}^*$ and $\kappa = j_1 \ldots j_m \in \{1, \ldots, n\}^*$. As usual, we treat the word κ as map from $\mathrm{dom}(\kappa) = \mathrm{dom}(v)$ to $\{1, \ldots, n\}$ with $\kappa(i) = j_i$ for any $i \in \mathrm{dom}(v)$. If $v \in \Sigma_{\mathcal{V}}^*$ is a valid word, i.e. $v \in N_{\mathcal{V}}$, then we write (κ, w, σ) for (κ, v), where (w, σ) is the pair obtained by decoding v. Further, we consider the language

$$\widetilde{L} = \{(\kappa, v) \in (\widetilde{\Sigma}_{\mathcal{V}})^* \mid v = (w, \sigma) \in N_{\mathcal{V}} \text{ and}$$
$$(w, \sigma[x \mapsto i]) \in L_{\kappa(i)} \text{ for any } i \in \mathrm{dom}(w)\}.$$

We observe that for each $(w, \sigma) \in N_{\mathcal{V}}$ there exists a unique word $\kappa \in \{1, \ldots, n\}^*$ such that $(\kappa, w, \sigma) \in \widetilde{L}$. Indeed, given a position $i \in \mathrm{dom}(w)$, the map $\sigma[x \mapsto i]$ is a $(\mathcal{W}, w)$–assignment, i.e. the encoding of $(w, \sigma[x \mapsto i])$ belongs to $\Sigma_{\mathcal{W}}^*$. Therefore, for any $i \in \mathrm{dom}(w)$ there exists a unique $j_i \in \{1, \ldots, n\}$ such that $(w, \sigma[x \mapsto i]) \in L_{j_i}$, since the languages $L_1, \ldots, L_n$ form a partition of $\Sigma_{\mathcal{W}}^*$. In particular, we then have $(\kappa, w, \sigma) \in \widetilde{L}$ for the uniquely determined $\kappa = j_1 \ldots j_{|w|} \in \{1, \ldots, n\}^{|w|}$.

Claim. *The language $\widetilde{L}$ over the extended alphabet $\widetilde{\Sigma}_{\mathcal{V}}$ is recognizable.*

Proof of Claim. First, let ψ be an arbitrary classical $\mathrm{MSO}(\Sigma)$–formula and $\mathcal{U} \subseteq \mathrm{Var}$ a finite set of variables containing $\mathrm{Free}(\psi)$. By replacing each atomic $\mathrm{MSO}(\Sigma)$–formula in ψ of the form $P_a(x)$ by the $\mathrm{MSO}(\widetilde{\Sigma})$–formula

$$\left(\bigvee_{j=1}^{n} P_{(j,a)}(x) \right),$$

we obtain an $\mathrm{MSO}(\widetilde{\Sigma})$–formula $\widetilde{\psi}$. This formula clearly fulfills $\mathrm{Free}(\widetilde{\psi}) = \mathrm{Free}(\psi) \subseteq \mathcal{U}$.

Moreover, it can be verified by structural induction that we have

$$(\kappa, w, \sigma) \models \widetilde{\psi} \quad (\text{in } \mathrm{MSO}(\widetilde{\Sigma}))$$
$$\Leftrightarrow (w, \sigma) \models \psi \quad (\text{in } \mathrm{MSO}(\Sigma))$$

for any word (κ, w, σ) over the extended alphabet $\widetilde{\Sigma}_{\mathcal{U}}$ with $(w, \sigma) \in N_{\mathcal{U}}$. For instance, we consider the crucial case $\psi = P_a(x)$. Given $(w, \sigma) \in N_{\mathcal{U}}$ and $\kappa \in \{1, \ldots, n\}^{|w|}$, we obtain

$$(\kappa, w, \sigma) \models \widetilde{\psi} \; \Leftrightarrow \; (\kappa, w, \sigma) \models \bigvee_{j=1}^{n} P_{(j,a)}(x)$$
$$\Leftrightarrow \; \kappa(\sigma(x)) \in \{1, \ldots, n\} \text{ and } w(\sigma(x)) = a$$
$$\Leftrightarrow \; w(\sigma(x)) = a$$
$$\Leftrightarrow \; (w, \sigma) \models P_a(x)$$
$$\Leftrightarrow \; (w, \sigma) \models \psi.$$

We consider now the recognizable languages $L_1, \ldots, L_n$ over the alphabet $\Sigma_{\mathcal{W}}$. By the generalized version of the Büchi–Elgot–Trakhtenbrot Theorem (see Theorem 2.5.19), for each $j \in \{1, \ldots, n\}$ there exists a classical $\mathrm{MSO}(\Sigma)$–formula ψ_j with $\mathrm{Free}(\psi_j) \subseteq \mathcal{W}$ such that

$$(w, \sigma) \models \psi_j \; \Leftrightarrow \; (w, \sigma) \in L_j$$

for any $(w, \sigma) \in N_{\mathcal{W}}$. If we define

$$\xi = \forall x \bigwedge_{j=1}^{n} \left(\left(\bigvee_{a \in \Sigma} P_{(j,a)}(x) \right) \to \widetilde{\psi_j} \right) \in \mathrm{MSO}(\widetilde{\Sigma}),$$

then we clearly have

$$\mathrm{Free}(\xi) \subseteq \mathcal{W} \setminus \{x\} = \mathcal{V},$$

as $\mathrm{Free}(\widetilde{\psi_j}) = \mathrm{Free}(\psi_j) \subseteq \mathcal{W}$ for any $j \in \{1, \ldots, n\}$. Moreover, considering an arbitrary word (κ, w, σ) over the extended alphabet $\widetilde{\Sigma}_{\mathcal{V}}$ with $(w, \sigma) \in N_{\mathcal{V}}$, our observations above yield

$$(\kappa, w, \sigma) \models \xi \; \Leftrightarrow \; \kappa(i) = j \text{ implies } (\kappa, w, \sigma[x \mapsto i]) \models \widetilde{\psi_j}$$
$$\text{for any } i \in \mathrm{dom}(w) \text{ and any } j \in \{1, \ldots, n\}$$
$$\Leftrightarrow \; (\kappa, w, \sigma[x \mapsto i]) \models \widetilde{\psi_{\kappa(i)}} \text{ for any } i \in \mathrm{dom}(w)$$
$$\Leftrightarrow \; (w, \sigma[x \mapsto i]) \models \psi_{\kappa(i)} \text{ for any } i \in \mathrm{dom}(w)$$
$$\Leftrightarrow \; (w, \sigma[x \mapsto i]) \in L_{\kappa(i)} \text{ for any } i \in \mathrm{dom}(w).$$

Therefore, we obtain $L_{\mathcal{V}}(\xi) = \widetilde{L}$, and hence $\widetilde{L}$ is recognizable by Proposition 2.5.12. $\diamond$

Since $\widetilde{L}$ is recognizable, there exists a deterministic finite automaton $\widetilde{\mathcal{A}} = (Q, I, T, F)$ over the extended alphabet $\widetilde{\Sigma}_{\mathcal{V}}$ such that $L(\widetilde{\mathcal{A}}) = \widetilde{L}$. We now define a weighted automaton $\mathcal{A} = (Q, \mathrm{in}, \mathrm{wt}, \mathrm{out})$ with the same state set by the weight functions

$$\mathrm{in}(q) = \begin{cases} 1 & \text{if } q \in I \\ 0 & \text{otherwise} \end{cases} \quad (q \in Q),$$

$$\mathrm{wt}(p, (j, a, \beta), q) = \begin{cases} s_j & \text{if } (p, (j, a, \beta), q) \in T \\ 0 & \text{otherwise} \end{cases} \quad (p, q \in Q, a \in \Sigma)$$

$$\mathrm{out}(q) = \begin{cases} 1 & \text{if } q \in F \\ 0 & \text{otherwise} \end{cases} \quad (q \in Q).$$

Since the classical automaton $\widetilde{\mathcal{A}}$ is deterministic, for each $W = (\kappa, w, \sigma) \in \widetilde{L}$ there is a unique successful path P_W in $\widetilde{\mathcal{A}}$ with label (κ, w, σ) and hence

$$(\|\mathcal{A}\|, (\kappa, w, \sigma)) = \sum_{\substack{P \in \mathrm{Path}(\mathcal{A}) \\ \mathrm{label}(P) = w}} \mathrm{weight}(P) = \mathrm{weight}(P_W),$$

whereas $(\|\mathcal{A}\|, (\kappa, v)) = 0$ for each $(\kappa, v) \in (\widetilde{\Sigma}_{\mathcal{V}})^* \setminus \widetilde{L}$. Given $W = (\kappa, w, \sigma) \in \widetilde{L}$, we write

$$P_W = q_0 (j_1, a_1, \beta_1) q_1 \ldots (j_{|w|}, a_{|w|}, \beta_{|w|}) q_{|w|}.$$

For any $i \in \mathrm{dom}(w)$ we then obtain

$$\mathrm{wt}(q_{i-1}, (j_i, a_i, \beta_i), q_i) = s_{j_i} = (\llbracket \varphi \rrbracket_{\mathcal{W}}, (w, \sigma[x \mapsto i])),$$

since $(\kappa, w, \sigma) \in \widetilde{L}$ implies $(w, \sigma[x \mapsto i]) \in L_{\kappa(i)} = L_{j_i}$, and the languages $L_1, \ldots, L_n$ form a partition of $\Sigma_{\mathcal{W}}^*$. This yields

$$(\|\mathcal{A}\|, (\kappa, w, \sigma)) = \mathrm{weight}(P_W) = \underbrace{\mathrm{in}(q_0)}_{\substack{=1 \\ \text{since } q_0 \in I}} \cdot \mathrm{rweight}(P_W) \cdot \underbrace{\mathrm{out}(q_{|w|})}_{\substack{=1 \\ \text{since } q_{|w|} \in F}}$$

$$= \mathrm{rweight}(P_W)$$

$$= \prod_{i \in \mathrm{dom}(w)} \mathrm{wt}(q_{i-1}, (j_i, a_i, \beta_i), q_i) = \prod_{i \in \mathrm{dom}(w)} s_{j_i}$$

$$= \prod_{i \in \mathrm{dom}(w)} \llbracket \varphi \rrbracket_{\mathcal{W}} (w, \sigma[x \mapsto i])$$

$$= \llbracket \forall x\, \varphi \rrbracket (w, \sigma).$$

In particular, considering the empty word $\varepsilon \in \widetilde{L}$, we have $(\|\mathcal{A}\|, \varepsilon) = 1 = \llbracket \forall x\, \varphi \rrbracket(\varepsilon)$. Finally, we denote by $h : (\widetilde{\Sigma}_{\mathcal{V}})^* \to \Sigma_{\mathcal{V}}^*$ the unique length-preserving monoid homomorphism extending the map

$$\widetilde{\Sigma}_{\mathcal{V}} \to \Sigma_{\mathcal{V}}, \quad (j, a, \beta) \mapsto (a, \beta).$$

As usual, we understand h as a projection, since it maps each word (κ, v) over the extended alphabet $\widetilde{\Sigma}_{\mathcal{V}}$ to the word v over $\Sigma_{\mathcal{V}}$. Then for any $(w, \sigma) \in N_{\mathcal{V}}$ and the unique $\kappa \in \{1, \ldots, n\}^*$ such that $(\kappa, w, \sigma) \in \widetilde{L}$, our observations above yield

$$(h(||\mathcal{A}||), (w, \sigma)) = \sum_{\substack{u \in h^{-1}(w, \sigma) \\ \underbrace{}_{=0} \\ \text{if } u \notin \widetilde{L}}} \underbrace{(||\mathcal{A}||, u)}$$

$$= (||\mathcal{A}||, (\kappa, w, \sigma))$$

$$= [\![\forall x\, \varphi]\!](w, \sigma),$$

using the transformation of formal power series induced by h (see Definition 3.2.13b)). On the other hand, given a non-valid word $v \in \Sigma_{\mathcal{V}}^*$, i.e. $v \notin N_{\mathcal{V}}$, any preimage u of v under the projection h does not belong to $\widetilde{L}$, i.e. , i.e. $u \in h^{-1}(v)$ implies $u \notin \widetilde{L}$, yielding

$$(h(||\mathcal{A}||), v) = \sum_{\substack{u \in h^{-1}(v) \\ \underbrace{}_{=0} \\ \text{since } u \notin \widetilde{L}}} \underbrace{(||\mathcal{A}||, u)} = 0 = [\![\forall x\, \varphi]\!](v).$$

Summarizing, we have

$$[\![\forall x\, \varphi]\!] = h(||\mathcal{A}||),$$

which is recognizable by Proposition 3.2.16. □

5.2.8 Theorem. *Let φ be a restricted $\mathrm{MSO}(\Sigma, S)$–sentence. If the semiring S is commutative, then the $\mathcal{V}$–semantics $[\![\varphi]\!]_{\mathcal{V}}$ is recognizable for any finite set $\mathcal{V} \subseteq \mathrm{Var}$ of variables containing $\mathrm{Free}(\varphi)$, i.e. $[\![\varphi]\!]_{\mathcal{V}} \in S^{\mathrm{rec}}\langle\!\langle \Sigma_{\mathcal{V}}^* \rangle\!\rangle$. In particular, we obtain*

$$S^{\mathrm{rmso}}\langle\!\langle \Sigma^* \rangle\!\rangle \subseteq S^{\mathrm{rec}}\langle\!\langle \Sigma^* \rangle\!\rangle.$$

Proof. Proceeding by structural induction and applying Proposition 5.2.3, Proposition 5.2.4, Proposition 5.2.5 and Proposition 5.2.7, yields that the series $[\![\varphi]\!] \in S\langle\!\langle \Sigma_{\mathcal{V}}^* \rangle\!\rangle$ is recognizable for any restricted $\mathrm{MSO}(\Sigma, S)$–formula φ, if the semiring S is commutative. The recognizability of the $\mathcal{V}$–semantics $[\![\varphi]\!]_{\mathcal{V}} \in S\langle\!\langle \Sigma_{\mathcal{V}}^* \rangle\!\rangle$ then follows directly from Lemma 5.2.2. In particular, the semantics of any sentence in the restricted logic $\mathrm{RMSO}(\Sigma, S)$ is recognizable, i.e. belongs to $S^{\mathrm{rec}}\langle\!\langle \Sigma^* \rangle\!\rangle$, which implies the inclusion

$$S^{\mathrm{rmso}}\langle\!\langle \Sigma^* \rangle\!\rangle \subseteq S^{\mathrm{rec}}\langle\!\langle \Sigma^* \rangle\!\rangle.$$

□

In [10, Theorem 5.6], Droste and Gastin extend the statement of Theorem 5.2.8 to arbitrary semirings, i.e. even to non-commutative ones.

Next, we aim at proving the inclusion

$$S^{\mathrm{rec}}\langle\!\langle \Sigma^* \rangle\!\rangle \subseteq S^{\mathrm{remso}}\langle\!\langle \Sigma^* \rangle\!\rangle.$$

From Weighted Automata to Weighted Formulas

To prove the inclusion $S^{rec}\langle\langle\Sigma^*\rangle\rangle \subseteq S^{remso}\langle\langle\Sigma^*\rangle\rangle$, we proceed similarly as in the classical setting (see Proposition 2.5.2). More precisely, given a weighted automaton $\mathcal{A}$, we construct an explicit weighted sentence $\varphi \in \text{REMSO}(\Sigma, S)$ such that $[\![\varphi]\!] = \|\mathcal{A}\|$. In this construction, the concept of *unambiguous* formulas, which is due to Droste and Gastin [6, Definition 5.1], will be helpful.

5.2.9 Definition. The set of **unambiguous** $\text{MSO}(\Sigma, S)$–formulas is defined inductively as follows:

(1) All literals in $\text{MSO}(\Sigma, S)$ are unambiguous.

(2) If φ and ψ are unambiguous $\text{MSO}(\Sigma, S)$–formulas, $x \in \text{Var}_1$ and $X \in \text{Var}_2$, then the conjunction $(\varphi \wedge \psi)$, and the universal quantifications $\forall x\,\varphi$ and $\forall X \varphi$ are again unambiguous.

(3) If φ and ψ are unambiguous $\text{MSO}(\Sigma, S)$–formulas such that

$$\text{supp}([\![\varphi]\!]) \cap \text{supp}([\![\psi]\!]) = \emptyset,$$

then the disjunction $\varphi \vee \psi$ is again unambiguous.

(4) Let φ be an unambiguous $\text{MSO}(\Sigma, S)$–formula, $x \in \text{Var}_1$, and set $\mathcal{V} = \text{Free}(\varphi)$. If for any $(w, \sigma) \in N_{\mathcal{V}}$ there is at most one position $i \in \text{dom}(w)$ such that

$$[\![\varphi]\!]_{\mathcal{V} \cup \{x\}}(w, \sigma[x \mapsto i]) \neq 0,$$

then also the first-order existential quantification $\exists x\,\varphi$ is unambiguous.

(5) Let φ be an unambiguous $\text{MSO}(\Sigma, S)$–formula, $X \in \text{Var}_2$, and set $\mathcal{V} = \text{Free}(\varphi)$. If for any $(w, \sigma) \in N_{\mathcal{V}}$ there is at most one subset $I \subseteq \text{dom}(w)$ such that

$$[\![\varphi]\!]_{\mathcal{V} \cup \{X\}}(w, \sigma[X \mapsto I]) \neq 0,$$

then also the second-order existential quantification $\exists X \varphi$ is unambiguous.

First, we observe that due to (3)–(5) this is not a purely syntactic definition, since the restriction of disjunctions and existential quantifications are on the semantics of formulas. As for the restricted formulas, Droste and Gastin later developed also a purely syntactical approach. More precisely, in [10] they consider the set of *syntactically unambiguous* $\text{MSO}(\Sigma, S)$–formulas. However, we follow their first approach in [5, § 5] and [6, § 5].

The notion of an *unambiguous* formula is reasonable, since the semantics of an unambiguous $\text{MSO}(\Sigma, S)$–formula is an unambiguous power series, i.e. the semantics takes only the values 0 and 1 in S. More precisely, for an unambiguous $\text{MSO}(\Sigma, S)$–formula the weighted semantics coincides with the classical Boolean semantics in the following sense:

5.2.10 Proposition. *Let φ be an unambiguous MSO(Σ, S)-formula. Then we can consider φ also as classical MSO(Σ)-formula to obtain*

$$\llbracket \varphi \rrbracket = \mathbb{1}_{L(\varphi)}.$$

In particular, the semantics of an unambiguous MSO(Σ, S)-formula is a recognizable step function.

Proof. For any non-valid word $v \in \Sigma_\varphi^*$ we have $\llbracket \varphi \rrbracket(v) = 0 = \mathbb{1}_{L(\varphi)}$, since $v \notin N_\varphi$ implies $v \notin L(\varphi)$. On the other hand, exploiting the unambiguity of φ, one can show by structural induction that we have

$$\llbracket \varphi \rrbracket(w, \sigma) = \begin{cases} 1 & \text{if } (w, \sigma) \models \varphi \\ 0 & \text{otherwise} \end{cases}$$

for any $(w, \sigma) \in N_\varphi$. We already encountered this identity for literals in Remark 5.1.5b). Summarizing, this yields $\llbracket \varphi \rrbracket = \mathbb{1}_{L(\varphi)}$, and since $L(\varphi)$ is a recognizable language by Proposition 2.5.12, the series $\llbracket \varphi \rrbracket$ is a recognizable step function. □

The next result, which is proved in [6, Lemma 5.3], shows that conversely certain classical MSO(Σ)-formulas can be effectively transformed into unambiguous MSO(Σ, S)-formulas, that are equivalent with respect to the Boolean semantics.

5.2.11 Lemma. *For each classical MSO(Σ)-formula φ not containing second-order quantifications (but possibly including atomic formulas of the form $x \in X$), we can effectively construct an unambiguous MSO(Σ, S)-formula $\varphi^\times$ such that $\llbracket \varphi^\times \rrbracket = \mathbb{1}_{L(\varphi)}$, i.e. for any $(w, \sigma) \in N_\varphi$ we have*

$$\llbracket \varphi^\times \rrbracket(w, \sigma) = 1 \iff (w, \sigma) \models \varphi.$$

5.2.12 Proposition. *For each classical MSO(Σ)-sentence φ we can effectively construct an unambiguous MSO(Σ, S)-sentence ψ such that $\llbracket \psi \rrbracket = \mathbb{1}_{L(\varphi)}$.*

Proof. First, we consider the language

$$\overline{L(\varphi)} = \Sigma^* \setminus L(\varphi) = \{w \in \Sigma^* \mid w \not\models \varphi\} = L(\neg\varphi).$$

By Corollary 2.5.14, the language $\overline{L(\varphi)} = L(\neg\varphi)$ is definable by some existential MSO(Σ)-sentence. In particular, we obtain

$$L(\varphi) = L(\forall X_1 \ldots \forall X_n\, \xi)$$

for some $n \in \mathbb{N}^+$ and an MSO(Σ)-formula ξ containing no second-order quantifications. Thus, if we set $\psi = \forall X_1 \ldots \forall X_n\, \xi^\times$, then Lemma 5.2.11 implies the claim. □

5.2.13 Remark. We observe that each unambiguous MSO(Σ, S)-formula φ that does not contain universal second-order quantifications is restricted. Indeed, if φ contains a subformula of the form $\forall x\, \psi$, then ψ is a again an unambiguous MSO(Σ, S)-formula, and thus its semantics $\llbracket \psi \rrbracket$ is a recognizable step function (see Proposition 5.2.10). This yields $\varphi \in \text{RMSO}(\Sigma, S)$.

Before we prove that each recognizable power series is also $\mathrm{REMSO}(\Sigma, S)$–definable, we first introduce some abbreviations that will help us in the proof.

5.2.14 Notation. Let $x \in \mathrm{Var}_1, X \in \mathrm{Var}_2$ and $s \in S$. We set

$$(x \in X \rightarrow s) := ((x \in X \wedge s) \vee \neg x \in X).$$

Given a finite set $\mathcal{V} \subseteq \mathrm{Var}$ of variables such that $\{x, X\} \subseteq \mathcal{V}$ and a $(w, \sigma) \in N_\mathcal{V}$, we then obtain

$$[\![x \in X \rightarrow s]\!]_\mathcal{V}(w, \sigma) = \begin{cases} s & \text{if } \sigma(x) \in \sigma(X) \\ 1 & \text{otherwise} \end{cases}$$

and $[\![x \in X \rightarrow s]\!]_\mathcal{V}(v)$ for any non-valid word $v \in \Sigma_\mathcal{V}^*$, i.e. for any $v \notin N_\mathcal{V}$. Thus,

$$[\![x \in X \rightarrow s]\!]_\mathcal{V} = s \cdot \mathbb{1}_{L_\mathcal{V}(x \in X)} + 1 \cdot \mathbb{1}_{L_\mathcal{V}(\neg x \in X)}$$

is a recognizable step function, as the languages $L_\mathcal{V}(x \in X)$ and $L_\mathcal{V}(\neg x \in X)$ are recognizable by Proposition 2.5.12. In particular, the $\mathrm{MSO}(\Sigma)$–formula $\forall x(x \in X \rightarrow s)$ is restricted and fulfills

$$[\![\forall x\,(x \in X \rightarrow s)]\!]_\mathcal{V}(w, \sigma) = \prod_{i \in \mathrm{dom}(w)} [\![x \in X \rightarrow s]\!]_{\mathcal{V} \cup \{x\}}(w, \sigma[x \mapsto i]) = s^{|\sigma(X)|}$$

for any $(w, \sigma) \in N_\mathcal{V}$.

5.2.15 Notation. For any $x, y, z \in \mathrm{Var}_1$, $X_1, \ldots, X_m \in \mathrm{Var}_2$ and $m \in \mathbb{N}$, we set:

- $\min(y) := \forall x\, y \leq x$

- $\max(z) := \forall x\, x \leq z$

- $\mathrm{suc}(x, y) := (x \leq y \wedge \neg y \leq x \wedge \forall z\,(z \leq x \vee y \leq z))$

- $\mathrm{partition}(X_0, \ldots, X_m) := \forall x \bigvee_{i=0}^{m} \left(x \in X_i \wedge \bigwedge_{\substack{j=0 \\ j \neq i}}^{m} \neg x \in X_j \right)$

By our observation in Remark 5.2.13, we obtain that $\min(y)$ and $\max(z)$ are restricted $\mathrm{MSO}(\Sigma, S)$–formulas, since they are both unambiguous and do not contain second-order universal quantifications. On the other hand, we understand $\mathrm{suc}(x, y)$ and $\mathrm{partition}(X_0, \ldots, X_m)$ as classical $\mathrm{MSO}(\Sigma)$–formulas in the following.

The following result is due to [6, Theorem 5.5]. For the proof, we however also took inspiration from [10, Proof of Theorem 5.7].

5.2.16 Theorem. *If the semiring S is commutative, then we have*

$$S^{\mathrm{rec}} \langle\!\langle \Sigma^* \rangle\!\rangle \subseteq S^{\mathrm{remso}} \langle\!\langle \Sigma^* \rangle\!\rangle,$$

i.e. any recognizable power series $r \in S\langle\!\langle \Sigma^* \rangle\!\rangle$ is $\mathrm{REMSO}(\Sigma, S)$–definable.

Proof. Let $\mathcal{A} = (Q, \text{in}, \text{wt}, \text{out})$ be a weighted automaton over Σ with weights in S. Our aim is to prove that there is a $\text{REMSO}(\Sigma, S)$–sentence whose semantics coincides with the behavior of $\mathcal{A}$, i.e. $\|\mathcal{A}\| \in S^{\text{remso}} \langle\!\langle \Sigma^* \rangle\!\rangle$. We consider the set

$$\mathcal{V} = \{X_{p,a,q} \mid p, q \in Q, a \in \Sigma\} \subseteq \text{Var}_2,$$

where $X_{p,a,q}$ denotes a second-order variable for each transition $(p, a, q) \in Q \times \Sigma \times Q$, and we choose an enumeration $\overline{X} = (X_0, \ldots, X_m)$ of $\mathcal{V}$ with $m = (|Q|^2 \cdot |A| - 1) \in \mathbb{N}$. First, we consider the $\text{MSO}(\Sigma, S)$–formula

$$\varphi_{\text{path}} := \text{partition}(\overline{X})^{\times} \wedge \underbrace{\bigwedge_{\substack{p,q\in Q \\ a\in\Sigma}} \forall x \, (x \in X_{p,a,q} \to P_a(x))^{\times}}_{=: \, \varphi_{\text{lab}}}$$

$$\wedge \, \forall x \, \forall y \, \underbrace{\left(\text{suc}(x, y) \to \bigvee_{\substack{p,q,r\in Q \\ a,b\in\Sigma}} (x \in X_{p,a,q} \wedge y \in X_{q,b,r}) \right)^{\times}}_{=: \, \varphi_{\text{trans}}} .$$

We have $\text{Free}(\varphi_{\text{path}}) = \mathcal{V}$, and φ_{path} is unambiguous by Lemma 5.2.11. Thus, by Proposition 5.2.10 its semantics $[\![\varphi_{\text{path}}]\!]$ is an unambiguous power series and therefore takes only the values 0 and 1 in S. Since φ_{path} does not contain any second-order quantifications, we further obtain that φ_{path} is restricted, i.e. $\varphi_{\text{path}} \in \text{RMSO}(\Sigma, S)$ (see Remark 5.2.13).

Claim. *Let $w = a_1 \ldots a_n \in \Sigma^+$ be non-empty. There is a bijective correspondence between the set of paths in $\mathcal{A}$ with label w and the set of $(\mathcal{V}, w)$–assignments σ such that $[\![\varphi_{\text{path}}]\!](w, \sigma) = 1$.*

Proof of Claim. Given a path $P = q_0 a_1 q_1 \ldots a_n q_n$ in $\mathcal{A}$ with label w, we define the $(\mathcal{V}, w)$–assignment σ_P by

$$\sigma_P(X_{p,a,q}) := \{i \in \text{dom}(w) \mid (q_{i-1}, a_i, q_i) = (p, a, q)\}$$

for any transition $(p, a, q) \in Q \times \Sigma \times Q$. Thus, given a position $i \in \text{dom}(w)$ and a transition (p, a, q) we have $i \in \sigma_P(X_{p,a,q})$ if and only if (p, a, q) is the transition in the path P that realizes the letter $w(i) = a_i$. It requires a simple computation to show that the pair (w, σ_P) satisfies φ_{path}, i.e. we have $[\![\varphi_{\text{path}}]\!](w, \sigma_P) = 1$. Conversely, let σ be a $(\mathcal{V}, w)$–assignment such that $[\![\varphi_{\text{path}}]\!](w, \sigma) = 1$. Due to the subformulas $\text{partition}(\overline{X})$ and φ_{lab}, for any position $i \in \text{dom}(w)$ there is a uniquely determined transition (p_i, a_i, q_i) such that $i \in \sigma(X_{p_i, a_i, q_i})$ and $a_i = w(i)$. Furthermore, the subformula φ_{trans} ensures that we obtain $p_{i+1} = q_i$ if $i < n$. Thus, setting $q_0 = p_1$, we obtain a unique path $P = q_0 a_1 q_1 \ldots a_n q_n$ in $\mathcal{A}$ with label w that fulfills $\sigma_P = \sigma$. $\diamond$

Next, we consider the $\text{MSO}(\Sigma, S)$–formula

$$\varphi_{\text{weight}} := \varphi_{\text{path}} \wedge \varphi_{\text{in}} \wedge \varphi_{\text{wt}} \wedge \varphi_{\text{out}}.$$

where the subformulas $\varphi_{\text{in}}, \varphi_{\text{wt}}$ and φ_{out} are defined as follows:

$$\varphi_{\text{in}} := \exists y \left(\min(y) \wedge \bigvee_{\substack{p,q \in Q \\ a \in \Sigma}} (y \in X_{p,a,q} \wedge \text{in}(p)) \right)$$

$$\varphi_{\text{wt}} := \bigwedge_{\substack{p,q \in Q \\ a \in \Sigma}} \forall x \left(x \in X_{p,a,q} \to \text{wt}(p,a,q) \right)$$

$$\varphi_{\text{out}} := \exists z \left(\max(z) \wedge \bigvee_{\substack{p,q \in Q \\ a \in \Sigma}} (z \in X_{p,a,q} \wedge \text{out}(q)) \right)$$

Then clearly $\text{Free}(\varphi_{\text{weight}}) = \mathcal{V}$. As each of the subformulas $\varphi_{\text{in}}, \varphi_{\text{wt}}, \varphi_{\text{out}}$ and φ_{path} is restricted (see Notation 5.2.14 and Notation 5.2.15), also their conjunction is restricted, i.e. $\varphi_{\text{weight}} \in \text{RMSO}(\Sigma, S)$. Furthermore, we obtain

$$[\![\varphi_{\text{weight}}]\!] = [\![\varphi_{\text{path}}]\!] \odot [\![\varphi_{\text{in}}]\!] \odot [\![\varphi_{\text{wt}}]\!] \odot [\![\varphi_{\text{out}}]\!].$$

In particular, for any $v \in \Sigma_{\mathcal{V}}^*$ we have $[\![\varphi_{\text{weight}}]\!](v) = 0$ if $[\![\varphi_{\text{path}}]\!](v) = 0$. On the other hand, if we consider a path $P = q_0 a_1 q_1 \ldots a_n q_n$ in $\mathcal{A}$ with label $w = a_1 \ldots a_n \in \Sigma^+$, and denote by σ_P the associated $(\mathcal{V}, w)$–assignment, then we have $[\![\varphi_{\text{path}}]\!](w, \sigma_P) = 1$. Hence, exploiting that σ_P is given by

$$\sigma_P(X_{p,a,q}) = \{ i \in \text{dom}(w) \mid (q_{i-1}, a_i, q_i) = (p,a,q) \},$$

we obtain

$$[\![\varphi_{\text{weight}}]\!](w, \sigma_P) = \underbrace{[\![\varphi_{\text{in}}]\!](w, \sigma_P)}_{=\text{in}(q_0)} \cdot [\![\varphi_{\text{wt}}]\!](w, \sigma_P) \cdot \underbrace{[\![\varphi_{\text{out}}]\!](w, \sigma_P)}_{=\text{out}(q_n)}$$

$$= \text{in}(q_0) \cdot \prod_{\substack{p,q \in Q \\ a \in \Sigma}} \text{wt}(p,a,q)^{|\sigma_P(X_{p,a,q})|} \cdot \text{out}(q_n)$$

$$= \text{in}(q_0) \cdot \text{wt}(q_0, a_1, q_1) \cdot \ldots \cdot \text{wt}(q_{n-1}, a_n, q_n) \cdot \text{out}(q_n)$$

$$= \text{weight}(P).$$

Considering the empty word yields

$$[\![\varphi_{\text{weight}}]\!](\varepsilon) = 0$$

due to the first-order existential quantifications in φ_{in} and φ_{out}. Hence, in order to deal with the empty word, we define another $\text{MSO}(\Sigma, S)$–sentence by

$$\varphi_{\text{empty}} := s_\varepsilon \wedge \forall x \, \neg x \le x$$

where $s_\varepsilon := (\|\mathcal{A}\|, \varepsilon) = \sum_{q \in Q} \text{in}(q) \cdot \text{out}(q)$. We have $\varphi_{\text{empty}} \in \text{RMSO}(\Sigma, S)$ and

$$[\![\varphi_{\text{empty}}]\!](w) = s_\varepsilon \cdot \underbrace{[\![\forall x \, \neg x \le x]\!](w)}_{=0} = 0$$

for any non-empty $w \in \Sigma^+$.

On the other hand, we obtain

$$\llbracket\varphi_{\text{empty}}\rrbracket(\varepsilon) = s_\varepsilon \cdot \underbrace{\llbracket\forall x\, \neg x \le x\rrbracket(\varepsilon)}_{=1} = s_\varepsilon = (||\mathcal{A}||, \varepsilon)$$

for the empty word (see Remark 5.1.5e)). As the formulas φ_{weight} and φ_{empty} both belong to $\text{RMSO}(\Sigma, S)$ and contain no second-order quantifications, the $\text{MSO}(\Sigma, S)$–sentence

$$\varphi := \exists X_0 \ldots X_m (\varphi_{\text{weight}} \vee \varphi_{\text{empty}})$$

is restricted and existential, i.e. $\varphi \in \text{REMSO}(\Sigma, S)$. Furthermore, our observations above imply

$$\llbracket\varphi\rrbracket(w) = \sum_{\substack{\sigma\ (\mathcal{V},w)\text{-assignm.} \\ \text{if } \llbracket\varphi_{\text{path}}\rrbracket(w,\sigma)=0}} \underbrace{\llbracket\varphi_{\text{weight}}\rrbracket(w,\sigma)}_{=0} + \underbrace{\llbracket\varphi_{\text{empty}}\rrbracket(w,\sigma)}_{=0}$$

$$= \sum_{\substack{\sigma\ (\mathcal{V},w)\text{-assignm.} \\ \llbracket\varphi_{\text{path}}\rrbracket(w,\sigma)=1}} \llbracket\varphi_{\text{weight}}\rrbracket(w,\sigma)$$

$$= \sum_{\substack{P\in\text{Path}(\mathcal{A}) \\ \text{label}(P)=w}} \llbracket\varphi_{\text{weight}}\rrbracket(w,\sigma_P)$$

$$= \sum_{\substack{P\in\text{Path}(\mathcal{A}) \\ \text{label}(P)=w}} \text{weight}(P) = (||\mathcal{A}||, w)$$

for any non-empty $w \in \Sigma^+$, and for the empty word we obtain

$$\llbracket\varphi\rrbracket(\varepsilon) = \underbrace{\llbracket\varphi_{\text{weight}}\rrbracket(\varepsilon)}_{=0} + \llbracket\varphi_{\text{empty}}\rrbracket(\varepsilon)$$

$$= \llbracket\varphi_{\text{empty}}\rrbracket(\varepsilon) = s_\varepsilon = (||\mathcal{A}||, \varepsilon).$$

This yields that $||\mathcal{A}|| = \llbracket\varphi\rrbracket$ is $\text{REMSO}(\Sigma, S)$–definable, i.e. $||\mathcal{A}|| \in S^{\text{remso}}\langle\!\langle\Sigma^*\rangle\!\rangle$. Hence, any recognizable formal power series is $\text{REMSO}(\Sigma, S)$–definable, which establishes the inclusion

$$S^{\text{rec}}\langle\!\langle\Sigma^*\rangle\!\rangle \subseteq S^{\text{remso}}\langle\!\langle\Sigma^*\rangle\!\rangle.$$

$\square$

5.2.17 Remark. In the proof of Theorem 5.2.16, we only exploit the commutativity of S to show that

$$\llbracket\varphi_{\text{wt}}\rrbracket(w,\sigma_P) = \prod_{i\in\text{dom}(w)} \text{wt}(q_{i-1}, a_i, q_i) = \text{rweight}(P).$$

However, a slight modification of the $\text{RMSO}(\Sigma, S)$–formula φ_{wt} guarantees that, even if the underlying semiring S is not commutative, the inclusion $S^{\text{rec}}\langle\!\langle\Sigma^*\rangle\!\rangle \subseteq S^{\text{remso}}\langle\!\langle\Sigma^*\rangle\!\rangle$ holds true.

More precisely, we replace φ_{wt} by the $\mathrm{MSO}(\Sigma, S)$–formula

$$\varphi'_{\mathrm{wt}} := \forall x \bigwedge_{\substack{p,q \in Q \\ a \in \Sigma}} (x \in X_{p,a,q} \to \mathrm{wt}(p,a,q)).$$

We first verify that the modified formula φ'_{wt} is still restricted. By our observation in Notation 5.2.14, the semantics $[\![x \in X_{p,a,q} \to \mathrm{wt}(p,a,q)]\!]$ is a recognizable step function for any transition (p,a,q). As the class of recognizable step functions is closed under Hadamard products (cf. [10, Lemma 2.3]), the semantics of the conjunction in φ'_{wt} is again a recognizable step function. Hence, we obtain $\varphi'_{\mathrm{wt}} \in \mathrm{RMSO}(\Sigma, S)$.

Now we consider a path $P = q_0 a_1 q_1 \ldots a_n q_n$ in $\mathcal{A}$ with label $w = a_1 \ldots a_n \in \Sigma^+$, and denote by σ_P the associated $(\mathcal{V}, w)$–assignment, which is given by

$$\sigma_P(X_{p,a,q}) = \{i \in \mathrm{dom}(w) \mid (q_{i-1}, a_i, q_i) = (p,a,q)\}.$$

We then have

$$[\![x \in X_{p,a,q} \to \mathrm{wt}(p,a,q)]\!]_{\mathcal{V} \cup \{x\}}(w, \sigma_P) = \begin{cases} \mathrm{wt}(p,a,q) & \text{if } i \in \sigma_P(X_{p,a,q}) \\ 1 & \text{otherwise,} \end{cases}$$

which yields

$$[\![\varphi'_{\mathrm{wt}}]\!](w, \sigma_P) = \prod_{i \in \mathrm{dom}(w)} \prod_{\substack{p,q \in Q \\ a \in \Sigma}} [\![x \in X_{p,a,q} \to \mathrm{wt}(p,a,q)]\!]_{\mathcal{V} \cup \{x\}}(w, \sigma_P(x \mapsto i))$$

$$= \prod_{i \in \mathrm{dom}(w)} \mathrm{wt}(q_{i-1}, a_i, q_i) = \mathrm{rweight}(P),$$

even if S is not commutative. For this modification we were inspired by the proof of [10, Theorem 5.7].

Proof of Theorem 5.2.1. Theorem 5.2.8 and Theorem 5.2.16 provide the inclusions

$$S^{\mathrm{rec}} \langle\!\langle \Sigma^* \rangle\!\rangle \subseteq S^{\mathrm{remso}} \langle\!\langle \Sigma^* \rangle\!\rangle \subseteq S^{\mathrm{rmso}} \langle\!\langle \Sigma^* \rangle\!\rangle \subseteq S^{\mathrm{rec}} \langle\!\langle \Sigma^* \rangle\!\rangle,$$

which immediately establish

$$S^{\mathrm{rec}} \langle\!\langle \Sigma^* \rangle\!\rangle = S^{\mathrm{rmso}} \langle\!\langle \Sigma^* \rangle\!\rangle = S^{\mathrm{remso}} \langle\!\langle \Sigma^* \rangle\!\rangle.$$

$\square$

Our observations in Remark 5.2.17 show that in Theorem 5.2.16 it is not necessary to suppose that the underlying semiring is commutative. However, for the proof of Theorem 5.2.8, more precisely for Proposition 5.2.4, certain commutativity assumptions are essential.

Further, we observe that the proof of Theorem 5.2.16 is constructive. More precisely, given a weighted automaton $\mathcal{A}$ over Σ with weights in S, we can effectively construct a weighted sentence $\varphi_{\mathcal{A}} \in \mathrm{REMSO}(\Sigma, S)$ such that $[\![\varphi_{\mathcal{A}}]\!] = \|\mathcal{A}\|$.

On the other hand, the proof of Theorem 5.2.8 is in general not completely constructive, since in Proposition 5.2.7 we may not know the form of the recognizable step function $\llbracket \varphi \rrbracket$ (cf. [10, page 193]). However, under certain assumptions on the underlying semiring S, for any restricted weighted formula $\varphi \in \mathrm{RMSO}(\Sigma, S)$ one can effectively construct a weighted automaton $\mathcal{A}_\varphi$ such that $||\mathcal{A}_\varphi|| = \llbracket \varphi \rrbracket$ (cf. [6, Proposition 4.6]). Therefore, some of the decidability properties of the classical logic $\mathrm{MSO}(\Sigma)$ (see Corollary 2.5.16) can be generalized also to the weighted logic $\mathrm{MSO}(\Sigma, S)$. For instance, we consider the Equivalence Problem for the restricted logic $\mathrm{RMSO}(\Sigma, S)$, i.e. the question whether we have $\llbracket \varphi \rrbracket = \llbracket \psi \rrbracket$ for given weighted sentences $\varphi, \psi \in \mathrm{RMSO}(\Sigma, S)$. One can show that, e.g. for the field of rationals $S = \mathbb{Q}$, the Equivalence Problem is decidable (cf. [6, Corollary 4.7]). In contrast, e.g. for the tropical semiring $S = \mathrm{Trop}$, the Equivalence Problem is undecidable (cf. [6, page 80]).

Locally Finite Semirings

In Section 5.1 we have presented examples showing that Theorem 5.2.1 does in general not hold for arbitrary weighted $\mathrm{MSO}(\Sigma, S)$–sentences. Therefore, we have considered the fragment $\mathrm{RMSO}(\Sigma, S)$ of restricted weighted formulas in the previous subsections. In contrast, we now wish to address the underlying semiring instead of restricting the formulas. More precisely, we introduce the notion of *locally finite* semirings, and we show that for this large class of semirings even unrestricted weighted formulas define recognizable power series. Thus, for any locally finite semiring S, we establish the equality

$$S^{\mathrm{rec}} \langle\!\langle \Sigma^* \rangle\!\rangle = S^{\mathrm{mso}} \langle\!\langle \Sigma^* \rangle\!\rangle.$$

The definitions and results are mainly due to Droste and Gastin [6, § 6].

5.2.18 Definition. Let S be a semiring.

a) Let S' be a subsemiring of S. We call S' **finitely generated** if there exists a finite set $M \subseteq S$ such that $\langle M \rangle = S'$, where $\langle M \rangle$ denotes the subsemiring of S generated by the set M.

b) If each finitely generated subsemiring S' of S is finite, i.e. $|S'| < \infty$, then the semiring S is called **locally finite**.

5.2.19 Example. In [10, page 195], one finds the following examples of locally finite semirings:

a) Clearly, every finite semiring is locally finite.

b) Any semiring that is induced by a Boolean algebra or, more generally, by a bounded distributive lattice is locally finite. For instance, the max-min semiring

$$(\overline{\mathbb{R}}_{\geq 0}, \max, \min, 0, \infty),$$

which is used in operations research for maximum capacity problems of networks, is locally finite.

c) Another example of a locally finite semiring is provided by the Łukasiewicz semiring $S = ([0, 1], \max, \otimes, 0, 1)$.

d) Examples of infinite but locally finite fields are provided by the algebraic closures of the finite fields $\mathbb{F}_p$ for any prime p.

If S is a locally finite semiring, then each recognizable series $r \in S^{\mathrm{rec}}\langle\!\langle \Sigma^* \rangle\!\rangle$ can be shown to be a recognizable step function (cf. [6, Lemma 6.1]). This implies in particular that first-order universal quantification does preserve recognizability if the underlying semiring is locally finite (see Proposition 5.2.7). Moreover, it can be shown that also second-order universal quantification does preserve recognizability if the underlying semiring is commutative and locally finite (cf. [6, § 6]). Thus, if S is commutative and locally finite, then we obtain

$$S^{\mathrm{mso}}\langle\!\langle \Sigma^* \rangle\!\rangle \subseteq S^{\mathrm{rec}}\langle\!\langle \Sigma^* \rangle\!\rangle,$$

i.e. we do not have to restrict the weighted logic $\mathrm{MSO}(\Sigma, S)$. To prove the inclusion

$$S^{\mathrm{rec}}\langle\!\langle \Sigma^* \rangle\!\rangle \subseteq S^{\mathrm{mso}}\langle\!\langle \Sigma^* \rangle\!\rangle,$$

we have two possibilities if the underlying semiring S is locally finite:

- Since $\mathrm{REMSO}(\Sigma, S)$ is a fragment of $\mathrm{MSO}(\Sigma, S)$, we immediately obtain

$$S^{\mathrm{rec}}\langle\!\langle \Sigma^* \rangle\!\rangle \subseteq S^{\mathrm{remso}}\langle\!\langle \Sigma^* \rangle\!\rangle \subseteq S^{\mathrm{mso}}\langle\!\langle \Sigma^* \rangle\!\rangle$$

  by Theorem 5.2.16.

- Alternatively, we can exploit that the semiring S is locally finite. As we have mentioned above, each recognizable power series $r \in S^{\mathrm{rec}}\langle\!\langle \Sigma^* \rangle\!\rangle$ is then of the form

$$r = \sum_{i=1}^{n} s_i \cdot \mathbb{1}_{L_i}$$

  for some $n \in \mathbb{N}$, scalars $s_1, \ldots, s_n \in S$ and recognizable languages $L_1, \ldots, L_n$ over Σ. By Theorem 2.5.1, there are classical $\mathrm{MSO}(\Sigma)$–sentences $\varphi_1, \ldots, \varphi_n$ such that $L(\varphi_i) = L_i$ for $i = 1, \ldots, n$. Thus, by Proposition 5.2.12 there are unambiguous $\mathrm{MSO}(\Sigma, S)$–sentences $\psi_1, \ldots, \psi_n$ such that $[\![\psi_i]\!] = \mathbb{1}_{L(\varphi_i)} = \mathbb{1}_{L_i}$ for $i = 1, \ldots, n$. Hence, we obtain

$$r = \sum_{i=1}^{n} s_i \cdot [\![\psi_i]\!] = \sum_{i=1}^{n} [\![s]\!] \odot [\![\psi_i]\!] = \left[\!\left[\bigvee_{i=1}^{n} (s_i \wedge \psi_i) \right]\!\right],$$

  which proves that r is $\mathrm{MSO}(\Sigma, S)$–definable, i.e. $r \in S^{\mathrm{mso}}\langle\!\langle \Sigma^* \rangle\!\rangle$. This yields

$$S^{\mathrm{rec}}\langle\!\langle \Sigma^* \rangle\!\rangle \subseteq S^{\mathrm{mso}}\langle\!\langle \Sigma^* \rangle\!\rangle.$$

Summarizing, our observations above establish the following result, which provides another characterization of recognizable formal power series. A complete proof can be found in [6, Theorem 6.4].

5.2.20 Theorem (Droste and Gastin). *Let S be commutative and locally finite and let $r \in S\langle\langle\Sigma^*\rangle\rangle$ be a formal power series. Then r is recognizable if and only if r is* MSO(Σ, S)*-definable. This means*

$$S^{\mathrm{rec}}\langle\langle\Sigma^*\rangle\rangle = S^{\mathrm{mso}}\langle\langle\Sigma^*\rangle\rangle.$$

Theorem 5.2.20 applies in particular to the Boolean semiring $\mathbb{B}$, which is clearly commutative and locally finite. By Proposition 3.2.7 and Proposition 5.1.10), the power series in $\mathbb{B}^{\mathrm{rec}}\langle\langle\Sigma^*\rangle\rangle$ and $\mathbb{B}^{\mathrm{mso}}\langle\langle\Sigma^*\rangle\rangle$ correspond precisely to the languages over Σ that are recognizable, respectively definable by some classical MSO(Σ)–sentence. Hence, Theorem 5.2.20 contains the classical result of Büchi, Elgot and Trakhtenbrot (see Theorem 2.5.1) as a very special case.

In [6, § 7], Droste and Gastin investigate weighted first-order logic and its connections to *aperiodic* power series. For suitable semirings, they prove a quantitative extension of the classical equivalence result between aperiodic and first-order definable languages. Aperiodic as well as star-free formal power series and their relations are introduced and investigated in Droste and Gastin [7]. Combining [7, Theorem 6.7] and [6, Theorem 7.11] yields a counterpart of Theorem 2.5.18 in the realm of formal power series for suitable semirings.

To conclude this chapter, we would like to refer to [10, § 9], where Droste and Gastin mention a range of applications of weighted automata and their weighted logic and give the corresponding references. Moreover, they present a collection of open problems regarding the weighted logic MSO(Σ, S).

6. Summary and Further Research

The purpose of this chapter is to give an overview of the main correspondences and equivalences we collected throughout this work. Moreover, we wish to give pointers to further research areas in the context of weighted automata, formal power series and weighted logic.

6.1. Summary

In the previous chapters we have aimed at investigating multiple tools for the modeling of languages and formal power series, respectively. Furthermore, we analyzed connections between certain languages and formal power series as well as relationships among the different modeling approaches by examining their expressive power.

Languages

In Chapter 2 the objects under consideration have been languages, and we have presented three expressively equivalent modeling approaches: finite automata, rational operations and monadic second-order logic. The three approaches have in common that they provide *finite* tools for the specification of possibly infinite languages. We have further derived that they all have the same expressive power. More precisely, we have considered Kleene's Theorem on the equivalence of recognizability and rationality in the context of languages, and we have established the Büchi–Elgot–Trakhtenbrot Theorem, which exhibits that the recognizable languages coincide with the languages definable by monadic second-order sentences. Overall, combining Theorem 2.3.6, Theorem 2.5.1 and Corollary 2.5.14 yields the following characterization of recognizable languages by means of rational operations and monadic second-order logic:

Theorem 1 (Kleene, Büchi, Elgot and Trakhtenbrot). *Let Σ be an alphabet and let L be a language over Σ. Then the following conditions are equivalent:*

(i) The language L is recognizable.

(ii) The language L is rational.

(iii) The language L is definable by some $\mathrm{MSO}(\Sigma)$–sentence.

(iv) The language L is definable by some existential $\mathrm{MSO}(\Sigma)$–sentence.

Each of the pairwise equivalences in Theorem 1 is accompanied by a constructive algorithm. Thus, we are given explicit procedures to interchange the tools used for the specification of a given language. Graphically, Theorem 1 can be displayed as follows:

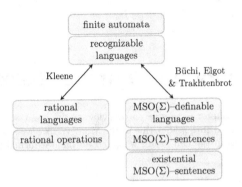

The Büchi–Elgot–Trakhtenbrot Theorem, as stated in Section 2.5, addresses only MSO(Σ)–sentences, i.e. MSO(Σ)–formulas without free variables. However, at the end of Section 2.5 we have derived a generalization, which also applies to MSO(Σ)–formulas having free variables (see Theorem 2.5.19).

Formal Power Series

In Section 3.1 we have then introduced formal power series. A formal power series is a map assigning to each word over Σ an element of a semiring S. Thus, formal power series provide quantitative information about words, more precisely associated weights or quantities, whereas languages provide only qualitative membership data. However, formal power series form a generalization of languages. More precisely, formal power series can be regarded as weighted, multi-valued or quantified languages in which each word comes with a weight, a number or some quantity. Via the injective map

$$\text{char}\colon \mathcal{P}(\Sigma^*) \to S\langle\!\langle \Sigma^* \rangle\!\rangle, \ L \mapsto \mathbb{1}_L$$

every language can be viewed as an unambiguous formal power series. Conversely, the map

$$\text{supp}\colon S\langle\!\langle \Sigma^* \rangle\!\rangle \to \mathcal{P}(\Sigma^*), \ r \mapsto \text{supp}(r)$$

assigns a language to each formal power series, but is in general not injective. However, the two maps provide a bijective correspondence between the set $\mathcal{P}(\Sigma^*)$ of languages and the set $\mathbb{B}\langle\!\langle \Sigma^* \rangle\!\rangle$ of formal power series with coefficients in the Boolean semiring $\mathbb{B}$:

$$\mathcal{P}(\Sigma^*) \ \underset{\text{supp}(r) \,\leftMapsfrom\, r}{\overset{L \,\mapsto\, \mathbb{1}_L}{\rightleftarrows}} \ \mathbb{B}\langle\!\langle \Sigma^* \rangle\!\rangle$$

Thus, the relation between languages and formal power series can be depicted as follows:

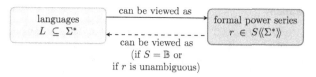

In the subsequent chapters, we have aimed at extending the language-theoretic equivalence results to the quantitative realm of formal power series. In particular, we have presented generalizations of the three representation tools from Chapter 2.

Weighted Automata

In Section 3.2 we have introduced the notion of weighted automata. While classical automata provide only qualitative membership information by deciding whether or not a word is recognized, weighted automata are capable of determining also quantitative properties. Formally, the behavior of a weighted automaton is represented by a formal power series with coefficients in a semiring S, and we have referred to such power series as *recognizable* power series. We have exhibited in Proposition 3.2.7 that the recognizable power series with coefficients in the Boolean semiring $\mathbb{B}$ correspond precisely to recognizable languages. In particular, we have shown that weighted automata are capable of capturing the qualitative method of recognition of classical automata. In Proposition 3.2.8 we have considered the relationship between recognizable languages and recognizable power series for arbitrary semirings. Graphically, we have established the following relationships:

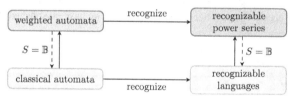

Linear Representations

In Section 3.3 we have represented weighted automata in terms of matrices. Hence, we have obtained a compact algebraic description for the behavior of weighted automata. We have further proved in Proposition 3.3.8 that these linear representations are expressively equivalent to weighted automata:

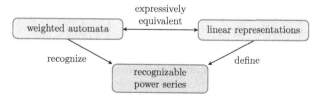

Rational Operations

In Section 4.1 and Section 4.2 we have aimed at extending the language-theoretic notions in the context of rationality to the realm of formal power series. In particular, we have defined operations on formal power series, and have drawn connections to the language-theoretic operations from Chapter 2 by considering the Boolean semiring. Overall, we have established multiple correspondences, which are summarized in the following tables:

basic components of rational ...	
... languages	... power series
singleton languages finite languages	monomials polynomials

operations on ...	
... languages	... power series
union intersection concatenation Kleene star	sum Hadamard product Cauchy product Kleene star / Neumann sum

In Proposition 4.2.3 we have exhibited that for the Boolean semiring $\mathbb{B}$ the notion of rational power series coincides with the one of rational languages. Moreover, we have considered the relationship between rational languages and rational power series for arbitrary semirings in Proposition 4.2.4. Graphically, we have established the following relationships:

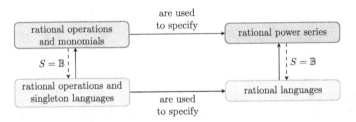

Weighted Monadic Second-Order Logic

In Section 5.1 we have introduced the syntax and semantics of the weighted monadic second-order logic $\mathrm{MSO}(\Sigma, S)$. By employing this formalism, we have obtained a further tool for the specification of formal power series. More precisely, the semantics of weighted monadic second-order formulas are represented by formal power series. Overall, in Chapter 5 we have aimed at establishing a generalization of the Büchi–Elgot–Trakhtenbrot in the realm of formal power series. In fact, we have exhibited in Proposition 5.1.10 that for the Boolean semiring $\mathbb{B}$ the weighted logic coincides with classical monadic second-order logic. Thus, we have drawn connections to the language-theoretic setting, which can be depicted as follows:

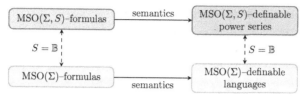

We showed that, as a consequence, one obtains $\mathbb{B}^{\mathrm{rec}}\langle\!\langle \Sigma^* \rangle\!\rangle = \mathbb{B}^{\mathrm{mso}}\langle\!\langle \Sigma^* \rangle\!\rangle = \mathbb{B}^{\mathrm{emso}}\langle\!\langle \Sigma^* \rangle\!\rangle$, i.e. that (existential) MSO($\Sigma, \mathbb{B}$)–sentences are expressively equivalent to weighted automata over Σ with weights in $\mathbb{B}$ (see Remark 5.1.22). However, weighted sentences over arbitrary semirings are in general too strong to capture the concept of recognizable power series. Therefore, we have introduced suitable fragments or restrictions of the weighted logic MSO(Σ, S). With the help of these, we have then derived logical characterizations of recognizable power series over commutative semirings.

Overall, we have presented four approaches for the specification of formal power series: weighted automata, linear representations, rational operations, and weighted monadic second-order logic. These approaches have in common that they provide *finite* tools for the representation of formal power series, which are in general infinite objects. Moreover, by outlining the investigations of Schützenberger as well as the joint work of Droste and Gastin we have derived that they are all expressively equivalent to each other. Hence, combining Proposition 3.3.8, Theorem 4.3.1, Theorem 5.2.1 and Theorem 5.2.1 yields the following characterizations of recognizable power series:

Theorem 2 (Schützenberger, Droste and Gastin). *Let Σ be an alphabet, S a nontrivial semiring, and let $r \in S\langle\!\langle \Sigma^* \rangle\!\rangle$ be a formal power series.*

a) The following conditions are equivalent:

  (i) The power series r is recognizable.

  (ii) There exists a linear representation (λ, μ, γ) such that $r = ||(\lambda, \mu, \gamma)||$.

  (iii) The power series r is rational.

b) If the semiring S is commutative, then the following conditions are equivalent:

  (i) The power series r is recognizable.

  (ii) The power series r is RMSO(Σ, S)–definable.

  (iii) The power series r is REMSO(Σ, S)–definable.

c) If the semiring S is commutative and locally finite, then the following conditions are equivalent:

  (i) The power series r is recognizable.

  (ii) The power series r is MSO(Σ, S)–definable.

In particular, we have

$$S^{\mathrm{rec}}\langle\!\langle\Sigma^*\rangle\!\rangle = S^{\mathrm{rat}}\langle\!\langle\Sigma^*\rangle\!\rangle = S^{\mathrm{rmso}}\langle\!\langle\Sigma^*\rangle\!\rangle = S^{\mathrm{remso}}\langle\!\langle\Sigma^*\rangle\!\rangle = S^{\mathrm{mso}}\langle\!\langle\Sigma^*\rangle\!\rangle$$

for any commutative and locally finite semiring S.

As in the language-theoretic setting, the proofs of the equivalences in Theorem 2 are accompanied by explicit procedures for interchanging the tools used for the specification of a given language. Graphically, Theorem 2 can be displayed as follows:

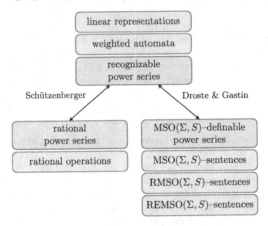

In the following example, we consider an explicit recognizable power series and specify all the tools given by the possible modeling approaches:

Example. Consider the alphabet $\Sigma = \{a, b\}$, the semiring $\mathbb{N}$ of natural numbers and the formal power series

$$r = \sum_{w \in \Sigma^*} |w|_a w \in \mathbb{N}\langle\!\langle\Sigma^*\rangle\!\rangle.$$

a) The power series r is recognizable, since it coincides with the behavior of the weighted automaton $\mathcal{A}$ which is represented by the following state diagram:

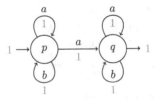

Indeed, we have seen in Example 3.2.11) that $r = ||\mathcal{A}||$.

Furthermore, in [2, Chapter I, Example 5.3] Berstel and Reutenauer further show that $r = ||(\lambda, \mu, \gamma)||$ for the linear representation (λ, μ, γ) of dimension $Q = \{p, q\}$ which is defined by the following matrices:

$$\lambda = \begin{pmatrix} \lambda_p & \lambda_q \end{pmatrix} = \begin{pmatrix} 1 & 0 \end{pmatrix},$$

$$\mu(a) = \begin{pmatrix} \mu(a)_{p,p} & \mu(a)_{p,q} \\ \mu(a)_{q,p} & \mu(a)_{q,q} \end{pmatrix} = \begin{pmatrix} 1 & 1 \\ 0 & 1 \end{pmatrix},$$

$$\mu(b) = \begin{pmatrix} \mu(b)_{p,p} & \mu(b)_{p,q} \\ \mu(b)_{q,p} & \mu(b)_{q,q} \end{pmatrix} = \begin{pmatrix} 1 & 0 \\ 0 & 1 \end{pmatrix},$$

$$\gamma = \begin{pmatrix} \gamma_p \\ \gamma_q \end{pmatrix} = \begin{pmatrix} 0 \\ 1 \end{pmatrix}.$$

b) The power series r is rational, since it can be written as

$$\sum_{w \in \Sigma^*} |w|_a w = \mathbb{1}_{\Sigma^*} \cdot 1a \cdot \mathbb{1}_{\Sigma^*}$$

$$= (\mathbb{1}_\Sigma)^* \cdot 1a \cdot (\mathbb{1}_\Sigma)^*$$

$$= (1a + 1b)^* \cdot 1a \cdot (1a + 1b)^*,$$

i.e. r can be constructed from the monomials $1a$ and $1b$ by application of the rational operations sum, Cauchy product and Kleene star (see Example 4.2.6).

c) The power series r is REMSO$(\Sigma, \mathbb{N})$–definable, and in particular MSO$(\Sigma, \mathbb{N})$–definable, since it coincides with the semantics of the REMSO$(\Sigma, \mathbb{N})$–sentence

$$\varphi = \exists x\, P_a(x).$$

Indeed, we have shown in Example 5.1.13 that $r = [\![\varphi]\!]$.

Ultimately, connecting Theorem 1 with Theorem 2 by the correspondences between the language-theoretic setting and the realm of formal power series yields the following diagram:

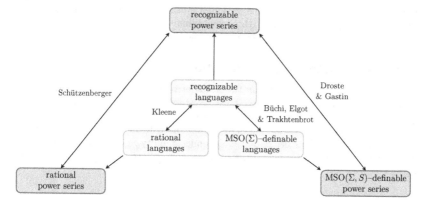

6.2. Further Research

The main focus of this work has been to state and prove quantitative extensions of Kleene's Theorem and the Büchi–Elgot–Trakhtenbrot Theorem for formal power series. However, there are many aspects of the theory of weighted automata, formal power series and weighted logic that we have touched on just slightly or not at all. Therefore, in this section we wish to give pointers to further research directions in the context of this theory.

Beyond Finite Words

Throughout this work, we have only dealt with finite words. However, another dimension of diversity evolves by considering (weighted) automata and (weighted) logics for discrete structures other than finite words, e.g. infinite words, trees, or pictures (cf. Droste, Kuich and Vogler [8, page vi]). For instance, weighted tree automata and transducers have been investigated e.g. for program analysis and transformation as well as for description logics, and weighted automata for infinite words are employed for image processing and are used as devices to compute real functions (cf. Droste and Kuske [12, page 142]). These automata models are further developed and investigated e.g. in [8, Part III]. Further references concerning the theory of (weighted) automata and (weighted) logics for discrete structures are given in [12, page 142] as well as in Droste and Gastin [6, page 85].

Beyond Free Monoids and Semirings

As we have focused on finite words, our study has been primarily concerned with the free monoid Σ^*. However, one can extend many of the results which we derived in this work by working with monoids in general, and not restricting them to the free monoid Σ^*. For details we refer to Sakarovitch [34]. Furthermore, we modeled the weight structures by semirings, and the distributivity of semirings allowed us to employ matrices, more precisely linear representations, for proofs regarding recognizable power series. However, in current research weighted automata over strong bimonoids have been investigated (cf. [12, page 141]). Strong bimonoids are algebraic structures that satisfy the same laws as semirings except that no distributivity laws need to hold.

Quantitative Automata and Valuations

Motivated by practical questions on the behavior of technical systems, new kinds of behaviors of weighted automata have been investigated. For instance, the run weight of a path could be the average of the transition weights. Moreover, one could be interested in discounting costs. Thus, valuations need to be included in the computation of weights. Various decidability and undecidability results, closure properties, and properties of the expressive powers of these models have been established (cf. [12, page 142]).

Applications

Finally, the general model of weighted automata and the theory of formal power series have found much interest in Computer Science due to its importance both in theory as well as in multiple practical applications. Indeed, weighted automata can be applied in a wide variety of situations, as the underlying weight structures are modeled by an arbitrary semiring. Weighted Automata have thus been employed as basic concepts in numerous practical applications such as natural language processing, speech recognition and digital image compression (cf. Droste, Kuich and Vogler [8, Part IV]). Since the early 1990s, weighted automata have been used for compressed representations of images and movies, which led to various algorithms for image transformation and processing (cf. Droste and Kuske [12, page 142 f.]). The use of weighted automata for digital image compression, in particular compared with the image compression standard JPEG, is presented in [8, Chapter 11]. The application of weighted automata in the context of probabilistic systems is highlighted in [8, Chapter 13]. In [8, Chapter 14] tasks occurring in natural language processing and speech recognition are concerned and solved by the use of weighted automata. Further special cases of weighted automata are networks with capacities, which have been investigated in operations research for algebraic optimization problems (cf. Droste and Gastin [7, page 609]). Weighted automata over De Morgan algebras form the basis for developments of practical tools for multi-valued model checking (cf. Droste and Kuske [12, page 143]). Treatments of further practical applications are provided by Pin [29, Part V].

For the wealth of theoretical developments of weighted automata and power series we refer to Eilenberg [14], Berstel and Reutenauer [2], Kuich and Salomaa [21], Salomaa and Soittola [35], Sakarovitch [33], Droste, Kuich and Vogler [8]. Each of these books deals with the interplay between weighted automata, semirings and formal power series. The books [14], [35] and [2] particularly deal with the theoretical aspects of the formal power series approach.

A. Appendix

The purpose of this chapter is to present the necessary background from Model Theory, which is a subarea of Mathematical Logic. The major concern of Model Theory is to study the relations between purely formal expressions on the syntactic level and their meaning on the semantic level. This relationship is established by means of interpretations in structures and a satisfaction relation. Basically, in this work we are mainly interested in interpreting logical formulas in words. Therefore, our treatment of the model-theoretic matters may be a bit unorthodox and somewhat "over-specialized", as we adapt them to our purposes.

First, Appendix A.1 provides an introduction to the general model-theoretic framework. In particular, we define the syntax and semantics of monadic second-order logic. In Appendix A.2 we then adapt the general model-theoretic notions from Appendix A.1 to our purpose of making logical statements about words. More precisely, we deal with a quite specialized framework of monadic second-order logic by considering a specific logical signature. Thereby, we obtain the formalism of monadic second-order logic for words, as treated in Section 2.4.

Our treatment is mainly based on Ebbinghaus, Flum and Thomas [13], Prestel and Delzell [30] and Libkin [22].

A.1. Model Theory and Monadic Second-Order Logic

A.1.1 Definition. A **signature** is a triple $\sigma = ((R_i)_{i \in I}, (f_j)_{j \in J}, (c_k)_{k \in K})$, where

- I, J and K are arbitrary index sets, which may even be empty,

- R_i is a **relation symbol** for any $i \in I$,

- f_j is a **function symbol** for any $j \in J$,

- c_k is a **constant symbol** for any $k \in K$,

- each relation and each function symbol has an associated **arity**. More precisely, the triple σ is implicitly provided with a function

$$\mathrm{ar} \colon I \cup J \to \mathbb{N}^+,$$

assigning to each $i \in I$ the arity $\mathrm{ar}(i) \in \mathbb{N}^+$ of the relation symbol R_i and to each $j \in J$ the arity $\mathrm{ar}(j) \in \mathbb{N}^+$ of the function symbol f_j. Usually, we write $\mathrm{ar}(R_i)$ for $\mathrm{ar}(i)$ and $\mathrm{ar}(f_j)$ for $\mathrm{ar}(j)$.

© The Editor(s) (if applicable) and The Author(s), under exclusive license to
Springer Fachmedien Wiesbaden GmbH, part of Springer Nature 2022
L. Wirth, *Weighted Automata, Formal Power Series and Weighted Logic*,
BestMasters, https://doi.org/10.1007/978-3-658-39323-6

Signatures are also referred to as **vocabularies** and relation symbols are also referred to as **predicate symbols**. A signature is called **purely relational** if the index sets J and K are empty, i.e. if σ just contains a family $(R_i)_{i \in I}$ of relation symbols. In this case, we simply write $\sigma = (R_i)_{i \in I}$. If we say that R is an n–ary relation symbol of σ, then we mean that $R = R_i$ for some $i \in I$ with $\mathrm{ar}(i) = \mathrm{ar}(R_i) = n$.

Let $\sigma = (R_i)_{i \in I}$ be a purely relational signature.

A more general treatment considering also arbitrary signatures can be found in [13], [30] and [22].

We now fix two countably infinite disjoint sets Var_1 and Var_2 of **first-order** and **second-order variables**, respectively, and we set $\mathrm{Var} := \mathrm{Var}_1 \cup \mathrm{Var}_2$. Occasionally, we refer to first-order variables as *individual variables* and to second-order variables as *set variables*. We typically denote individual variables by small letters like x, y, z, and set variables by capital letters like X, Y, Z, possibly with subscripts or superscripts.

To define logical formulas over a signature σ, one usually starts by defining the set of terms over σ (cf. [13, § II.3], [30, § 1.2]). However, the only terms with respect to relational signatures are first-order variables.

A.1.2 Definition. We define the set of **first-order formulas** over σ, or the set of FO(σ)–formulas, inductively as follows:

(1) $\bot$ is an FO(σ)–formula.

(2) If R is an n–ary relation symbol of σ and $x_1, \dots, x_n \in \mathrm{Var}_1$, then $R(x_1, \dots, x_n)$ is an FO(σ)–formula.

(3) If φ is an FO(σ)–formula, then so is its **negation** $\neg\varphi$.

(4) If φ and ψ are FO(σ)–formulas, then so is their **disjunction** $(\varphi \vee \psi)$.

(5) If φ is an FO(σ)–formula and $x \in \mathrm{Var}_1$, then the **first-order existential quantification** $\exists x\, \varphi$ is again an FO(σ)–formula.

We write $\varphi \in \mathrm{FO}(\sigma)$ if φ is an FO(σ)–formula. Moreover, we call the formulas obtained by steps (1) and (2) **atomic**.

A.1.3 Definition. We define the set of **monadic second-order formulas** over σ, or the set of MSO(σ)–formulas, inductively as follows:

(1) $\bot$ is an MSO(σ)–formula.

(2) If R is an n–ary relation symbol of σ and $x_1, \dots, x_n \in \mathrm{Var}_1$, then $R(x_1, \dots, x_n)$ is an MSO(σ)–formula.

(3) $x \in X$ is an MSO(σ)–formula for each $x \in \mathrm{Var}_1$ and $X \in \mathrm{Var}_2$.

(4) If φ is an MSO(σ)–formula, then so is its **negation** $\neg\varphi$.

(5) If φ and ψ are MSO(σ)–formulas, then so is their **disjunction** $(\varphi \vee \psi)$.

(6) If φ is an MSO(σ)–formula and $x \in \mathrm{Var}_1$, then the **first-order existential quantification** $\exists x \, \varphi$ is again an MSO(σ)–formula.

(7) If φ is an MSO(σ)–formula and $X \in \mathrm{Var}_2$, then the **second-order existential quantification** $\exists X \varphi$ is again an MSO(σ)–formula.

We write $\varphi \in \mathrm{MSO}(\sigma)$ if φ is an MSO(σ)–formula. Moreover, we call the formulas obtained by steps (1)–(3) **atomic**. Usually, we denote MSO(Σ)–formulas by Greek letters such as φ or ψ.

One can verify inductively that each FO(σ)–formula is also an MSO(σ)–formula, i.e. we have $\mathrm{FO}(\sigma) \subseteq \mathrm{MSO}(\sigma)$.

In this work, we only treat *monadic* second-order logic, which is a fragment of second-order logic. Therefore, we only consider second-order variables of arity 1, i.e. unary or *monadic* second-order variables, which will later be interpreted as sets. In particular, quantification is restricted to set variables. By defining second-order variables of arbitrary arity and also allowing quantification over these predicates, one obtains the general definition of second-order formulas.

A.1.4 Notation. Given a binary relation symbol R and two individual variables x, y, we usually write

- xRy instead of $R(x, y)$,

- $x\mathcal{R}y$ instead of $R(y, x)$,

- $x\not\!Ry$ instead of $\neg xRy$.

Next, we list some standard logical expressions and abbreviations.

A.1.5 Notation. We usually write

- $\top$ for $\neg\bot$,

- $x \notin X$ for $\neg x \in X$,

- $(\varphi \wedge \psi)$ for $\neg(\neg\varphi \vee \neg\psi)$, called **conjunction**,

- $(\varphi \to \psi)$ for $(\neg\varphi \vee \psi)$, called **implication**,

- $(\varphi \leftrightarrow \psi)$ for $((\varphi \to \psi) \wedge (\psi \to \varphi))$, called **(logical) equivalence**,

- $\forall x \, \varphi$ for $\neg\exists x \, \neg\varphi$, called **first-order universal quantification**,

- $\forall X \varphi$ for $\neg\exists X \neg\varphi$, called **second-order universal quantification**,

- $\exists x_1, x_2, \ldots, x_n \, \varphi$ for $\exists x_1 \exists x_2 \ldots \exists x_n \, \varphi$,

- $\forall x_1, x_2, \ldots, x_n \, \varphi$ for $\forall x_1 \forall x_2 \ldots \forall x_n \, \varphi$,

- $\exists X_1, X_2, \ldots, X_n \varphi$ for $\exists X_1 \exists X_2 \ldots \exists X_n \varphi$,

- $\forall X_1, X_2, \ldots, X_n \varphi$ for $\forall X_1 \forall X_2 \ldots \forall X_n \varphi$,

169

Moreover, we usually omit the outermost brackets when this cannot lead to any ambiguity, i.e. for instance, we write $\varphi \wedge \forall x\, (\psi_1 \vee \psi_2)$ instead of $(\varphi \wedge \forall x\, (\psi_1 \vee \psi_2))$.

A.1.6 Remark. Typically, a binary identity symbol $=$ is implicitly included in every signature (cf. [13, § II.3], [30, § 1.2], [22, § 2.1]). However, we refrain from doing so for two reasons. First, the equality $x = y$ of two individuals x, y can be expressed by the MSO(σ)–formula $\forall X\, (x \in X \leftrightarrow y \in X)$ (see Example A.1.14). Secondly, in Appendix A.2 we will deal with a signature containing the binary relation symbol $\leq$, and then $x = y$ can be expressed by the conjunction $(x \leq y \wedge y \leq x)$ (see Remark A.2.7).

To define the semantics of formulas, we first have to fix the boundaries of a domain of objects to which our quantifiers should refer, i.e. over which the variables should "vary" (cf. [30, page 36]). Furthermore, we have to specify interpretations of the relation symbols in the signature σ.

A.1.7 Definition. A σ–**structure** is a pair $\mathfrak{A} = (A, (R_i^{\mathfrak{A}})_{i \in I})$, where

- A is a set, which may be empty and is called **domain** or **universe** of $\mathfrak{A}$,

- $R_i^{\mathfrak{A}}$ is a relation on A of arity $\mathrm{ar}(R_i)$, called **interpretation** of the relation symbol R_i. More precisely, if R is an n–ary relation symbol of σ, then its interpretation $R^{\mathfrak{A}}$ is a subset of A^n.

If the domain of $\mathfrak{A}$ is empty, then the σ–structure $\mathfrak{A}$ is itself called empty and is denoted by $\emptyset$. For convenience, $R^{\mathfrak{A}}$ is usually written R^A or just R if the σ–structure is understood. The universe of a structure is typically denoted by a Roman letter corresponding to the name of the structure. Thus, e.g. the universe of the structure $\mathfrak{B}$ is denoted by B. Occasionally, both the structure as well as its domain are denoted by the same letter.

Typically, the empty σ–structure is excluded in Model Theory (cf. [13, § III.1], [30, § 1.5]). However, in Section 2.4 we represent words over a given alphabet as structures over a suitable signature, and in doing so the empty structure corresponds to the empty word ε (see Definition 2.4.4).

To formally define the semantics of MSO(σ)–formulas, we first need to specify *free occurrences* of variables in a formula.

A.1.8 Definition. Given $\varphi \in \mathrm{MSO}(\sigma)$, we define the set $\mathrm{Free}(\varphi)$ of **free variables** of the formula φ by structural induction as follows:

$$\mathrm{Free}(\bot) := \emptyset,$$
$$\mathrm{Free}(R(x_1, \ldots, x_n)) := \{x_1, \ldots, x_n\},$$
$$\mathrm{Free}(x \in X) := \{x, X\},$$
$$\mathrm{Free}(\neg \varphi) := \mathrm{Free}(\varphi),$$
$$\mathrm{Free}(\varphi \vee \psi) := (\mathrm{Free}(\varphi) \cup \mathrm{Free}(\psi)),$$
$$\mathrm{Free}(\exists x\, \varphi) := \mathrm{Free}(\varphi) \setminus \{x\},$$
$$\mathrm{Free}(\exists X \varphi) := \mathrm{Free}(\varphi) \setminus \{X\}.$$

A formula $\varphi \in \mathrm{MSO}(\sigma)$ with $\mathrm{Free}(\varphi) = \emptyset$ is called an $\mathrm{MSO}(\sigma)$–**sentence**.

A.1.9 Example. We consider a signature σ containing a binary relation symbol R. For the $\mathrm{MSO}(\sigma)$–formula $\varphi := \forall z\,(xRx \vee \exists x\;xRy \vee \forall z\;z \in X)$ we then obtain

$$\mathrm{Free}(\varphi) = \Big(\underbrace{\mathrm{Free}(xRx)}_{=\{x\}} \cup \underbrace{\mathrm{Free}(\exists x\;xRy)}_{=\{y\}} \cup \underbrace{\mathrm{Free}(\forall z\;z \in X)}_{=\{X\}} \Big) \setminus \{z\} = \{x, y, X\}.$$

Note that the variable x occurs various times. Although its occurrence in the subformula $\exists x\;xRy$ is *bound*, x is a free variable of the formula φ, since its occurrence in the subformula xRx is free. In general, the set $\mathrm{Free}(\varphi)$ consists precisely of those variables that possess at least one free occurrence in φ.

Without loss of generality, we assume that distinct quantifiers within a formula use pairwise distinct variables, and that no variable has both free and bound occurrences. For instance, we do not consider formulas like φ from Example A.1.9, since this formula can be replaced by the clearer formula

$$\varphi' := \forall z\,(xRx \vee \exists x'\;x'Ry \vee \forall z'\;z' \in X)$$

with the same meaning.

A.1.10 Definition. Let $\mathfrak{A}$ be a σ–structure and $\mathcal{V} \subseteq \mathrm{Var}$ be a finite set of variables.

a) We set $\mathcal{V}_i := (\mathcal{V} \cap \mathrm{Var}_i)$ for $i = 1, 2$. Thus, $\mathcal{V}_1$ contains all individual variables in $\mathcal{V}$ and $\mathcal{V}_2$ contains all set variables in $\mathcal{V}$.

b) A $(\mathcal{V}, \mathfrak{A})$–**assignment** is a map $\tau \colon \mathcal{V} \to A \cup \mathcal{P}(A)$ such that

$$\tau(\mathcal{V}_1) \subseteq A \quad \text{and}$$
$$\tau(\mathcal{V}_2) \subseteq \mathcal{P}(A).$$

By this, a $(\mathcal{V}, \mathfrak{A})$–assignment maps first-order variables in $\mathcal{V}$ to elements and second-order variables in $\mathcal{V}$ to subsets of the universe A of $\mathfrak{A}$.

c) Given a $(\mathcal{V}, \mathfrak{A})$–assignment τ, an element $a \in A$ and a first-order variable $x \in \mathrm{Var}_1$, we let $\tau[x \mapsto a]$ be the $(\mathcal{V} \cup \{x\}, \mathfrak{A})$–assignment mapping x to a and acting like τ elsewhere, i.e. $\tau[x \mapsto a]$ satisfies

$$\tau[x \mapsto a](x) = a \quad \text{and}$$
$$\tau[x \mapsto a]\big|_{\mathcal{V} \setminus \{x\}} = \tau\big|_{\mathcal{V} \setminus \{x\}}.$$

Similarly, if $X \in \mathrm{Var}_2$ is a second-order variable and $A' \subseteq A$, then we define a $(\mathcal{V} \cup \{X\}, \mathfrak{A})$–assignment $\tau[X \mapsto A']$ by

$$\tau[X \mapsto A'](X) = A' \quad \text{and}$$
$$\tau[X \mapsto A']\big|_{\mathcal{V} \setminus \{X\}} = \tau\big|_{\mathcal{V} \setminus \{X\}}.$$

The approach of restricting variable assignments to finite sets of variables has the advantage of allowing us to treat the structures in which we interpret formulas in Appendix A.2 and Section 2.4 as finite words over an extended alphabet (see Definition 2.5.3). For the purely model-theoretic use, one usually considers variable assignments having as domain the whole set Var of variables (cf. [13, § III,1] and [30, Appendix B]).

A.1.11 Remark. Let $\mathfrak{A}$ be a σ–structure, $\mathcal{V} \subseteq$ Var be a finite set of variables, τ a $(\mathcal{V}, \mathfrak{A})$–assignment, $x \in \mathcal{V}_1$ a first-order variable in $\mathcal{V}$, and $a \in A$ an element of the universe of $\mathfrak{A}$. Then we have $\mathcal{V} \cup \{x\} = \mathcal{V}$, and hence we have to distinguish two cases to evaluate the $(\mathcal{V}, \mathfrak{A})$–assignment $\tau[x \mapsto a]$ at the variable x: Either we have $\tau(x) = a$ and thus $\tau[x \mapsto a] = \tau$, or we have $\tau(x) \neq a$ and thus $\tau[x \mapsto a] \neq \tau$. More precisely, in the second case we obtain

$$\tau[x \mapsto a](x) = a \neq \tau(x) \text{ and } \tau[x \mapsto a]\big|_{\mathcal{V} \setminus \{x\}} = \tau\big|_{\mathcal{V} \setminus \{x\}}.$$

The same applies for $(\mathcal{V}, \mathfrak{A})$–assignments of the form $\tau[X \mapsto A']$ where $X \in \mathcal{V}_2$ is a second-order variable in $\mathcal{V}$ and $A' \in \mathcal{P}(A)$ is a subset of the universe of $\mathfrak{A}$.

One obtains the definition of *weak monadic second-order logic* if the variable assignments are restricted in the sense that set variables are assigned only *finite* subsets of the universe (cf. [13, § X.9]).

We are now ready to define the semantics of $\mathrm{MSO}(\sigma)$.

A.1.12 Definition. Let φ be an $\mathrm{MSO}(\sigma)$–formula, $\mathcal{V} \subseteq$ Var a finite set of variables such that $\mathrm{Free}(\varphi) \subseteq \mathcal{V}$, $\mathfrak{A}$ a non-empty σ–structure and τ a $(\mathcal{V}, \mathfrak{A})$–assignment. We define the **satisfaction relation** $(\mathfrak{A}, \tau) \models \varphi$, read $(\mathfrak{A}, \tau)$ **models** or **satisfies** φ, by structural induction as follows:

The atomic $\mathrm{MSO}(\sigma)$–formula $\bot$ is not fulfilled by any σ–structure, i.e. we unconditionally have

$$(\mathfrak{A}, \tau) \not\models \bot.$$

For the remaining cases, the satisfaction relation $(\mathfrak{A}, \tau) \models \varphi$ is defined as follows:

$$(\mathfrak{A}, \tau) \models R(x_1, \ldots, x_n) :\Leftrightarrow (\tau(x_1), \ldots, \tau(x_n)) \in R^{\mathfrak{A}}$$
$$(\mathfrak{A}, \tau) \models x \in X :\Leftrightarrow \tau(x) \in \tau(X),$$
$$(\mathfrak{A}, \tau) \models \neg\varphi :\Leftrightarrow (\mathfrak{A}, \tau) \not\models \varphi :\Leftrightarrow \text{not } (\mathfrak{A}, \tau) \models \varphi,$$
$$(\mathfrak{A}, \tau) \models (\varphi_1 \vee \varphi_2) :\Leftrightarrow (\mathfrak{A}, \tau) \models \varphi_1 \text{ or } (\mathfrak{A}, \tau) \models \varphi_2,$$
$$(\mathfrak{A}, \tau) \models \exists x\, \varphi :\Leftrightarrow (\mathfrak{A}, \tau[x \mapsto a]) \models \varphi \text{ for some } a \in A,$$
$$(\mathfrak{A}, \tau) \models \exists X \varphi :\Leftrightarrow (\mathfrak{A}, \tau[X \mapsto A']) \models \varphi \text{ for some } A' \subseteq A.$$

The semantics for the empty σ–structure $\mathfrak{A} = \emptyset$ are slightly different. The empty structure satisfies all atomic $\mathrm{MSO}(\Sigma)$–formulas except $\bot$, i.e. the relations

$$\emptyset \not\models \bot,$$
$$\emptyset \models R(x_1, \ldots, x_n),$$
$$\emptyset \models x \in X$$

hold unconditionally.

On the contrary, first-order existential quantifications are never satisfied by the empty structure, i.e. we unconditionally have

$$\emptyset \not\models \exists x\, \varphi.$$

The satisfaction of negations and disjunctions is defined as usual:

$$\emptyset \models \neg\varphi \;:\Leftrightarrow\; \emptyset \not\models \varphi \;:\Leftrightarrow\; \text{not } \emptyset \models \varphi,$$

$$\emptyset \models (\varphi_1 \vee \varphi_2) \;:\Leftrightarrow\; \emptyset \models \varphi_1 \text{ or } \emptyset \models \varphi_2.$$

Further, we set

$$\emptyset \models \exists X\varphi \;:\Leftrightarrow\; \emptyset \models \varphi.$$

A.1.13 Remark.

a) In Definition A.1.12, the word "or" is to be understood in the usual mathematical sense, i.e. it is not used in the exclusive sense (cf. [13, § III.2]).

b) It clearly makes sense to say that the empty structure never satisfies first-order existential quantifications. As a consequence, the empty structure does satisfy any first-order universal quantification. In Model Theory, one usually says that a structure $\mathfrak{A}$ satisfies a formula φ, in symbols $\mathfrak{A} \models \varphi$, if it satisfies the *universal closure* of φ (see Definition A.1.16 and Remark A.1.17). As the universal closure of a formula with free first-order variables is a first-order universal quantification, it makes sense to define that the empty structure unconditionally satisfies all atomic formulas. Since the only subset of the empty set is the empty set itself, we should have

$$\emptyset \models \exists X\varphi \;\Leftrightarrow\; \emptyset \models \forall X\varphi.$$

Indeed, by Definition A.1.12 we obtain

$$\emptyset \models \exists X\varphi \;\Leftrightarrow\; \emptyset \models \varphi \;\Leftrightarrow\; \emptyset \not\models \neg\varphi \;\Leftrightarrow\; \emptyset \not\models \exists X\varphi \;\Leftrightarrow\; \emptyset \models \neg\exists X\varphi \;\Leftrightarrow\; \emptyset \models \forall X\varphi.$$

A.1.14 Example. Consider any σ–structure $\mathfrak{A}$ and the MSO(σ)–formula

$$\varphi := \forall X (x \in X \leftrightarrow y \in X),$$

set $\mathcal{V} = \text{Free}(\varphi) = \{x, y\}$, and let τ be a $(\mathcal{V}, \mathfrak{A})$–assignment. Then we have

$$(\mathfrak{A}, \tau) \models \varphi \;\Leftrightarrow\; \tau(x) = \tau(y).$$

Indeed, $(\mathfrak{A}, \tau) \models \varphi$ implies in particular that we have

$$\tau(x) \in \{\tau(x)\} \text{ if and only if } \tau(y) \in \{\tau(x)\},$$

since $\{\tau(x)\}$ is a subset of the universe of $\mathfrak{A}$. This obviously yields $\tau(x) = \tau(y)$. Conversely, assuming that $\tau(x) = \tau(y)$, we clearly obtain

$$\tau(x) \in A' \text{ if and only if } \tau(y) \in A'$$

for any subset A' of the universe of $\mathfrak{A}$, which implies $(\mathfrak{A}, \tau) \models \varphi$. Hence, the MSO($\Sigma$)–formula φ expresses the equality of two individuals.

A.1.15 Lemma. *Let φ be an $\mathrm{MSO}(\sigma)$-formula, $\mathcal{V} \subseteq \mathrm{Var}$ a finite set of variables such that $\mathrm{Free}(\varphi) \subseteq \mathcal{V}$, $\mathfrak{A}$ a non-empty σ-structure and τ a $(\mathcal{V}, \mathfrak{A})$-assignment. We have*

$$(\mathfrak{A}, \tau) \models \varphi \; \Leftrightarrow \; (\mathfrak{A}, \tau|_{\mathrm{Free}(\varphi)}) \models \varphi$$

i.e. the satisfaction of φ only depends on $\mathfrak{A}$ and the restriction $\tau|_{\mathrm{Free}(\varphi)}$ of τ to $\mathrm{Free}(\varphi)$.

Proof. If φ is an $\mathrm{FO}(\sigma)$-formula, then the claim follows immediately from Prestel and Delzell [30, Lemma 1.5.1]. One can proceed similarly to prove the claim also for $\mathrm{MSO}(\sigma)$-formulas. $\qquad\square$

Hence, whether or not we have $(\mathfrak{A}, \tau) \models \varphi$ only depends on the σ-structure $\mathfrak{A}$ and the evaluations of free variables in φ under the assignment τ. As sentences do not possess free variables, Lemma A.1.15 shows in particular that the satisfaction of a sentence does not depend on any variable assignment. More precisely, if φ is an $\mathrm{MSO}(\sigma)$-sentence, $\mathcal{V}, \mathcal{V}' \subseteq \mathrm{Var}$ are finite sets of variables, $\mathfrak{A}$ is a non-empty σ-structure, τ is a $(\mathcal{V}, \mathfrak{A})$-assignment and τ' is a $(\mathcal{V}', \mathfrak{A})$-assignment, then we have

$$(\mathfrak{A}, \tau) \models \varphi \; \Leftrightarrow \; (\mathfrak{A}, \tau') \models \varphi.$$

Therefore, we simply set

$$\mathfrak{A} \models \varphi \; :\Leftrightarrow \; (\mathfrak{A}, \epsilon) \models \varphi$$

where ϵ denotes the empty $(\mathrm{Free}(\varphi), \mathfrak{A})$-assignment, i.e. the empty assignment from $\mathrm{Free}(\varphi) = \emptyset$ into the union $A \cup \mathcal{P}(A)$, where A denotes the universe of $\mathfrak{A}$.

The following syntactic operation transform an $\mathrm{MSO}(\sigma)$-formula φ into an $\mathrm{MSO}(\sigma)$-sentence (cf. [30, page 13]).

A.1.16 Definition. Let φ be an $\mathrm{MSO}(\sigma)$-formula. If the set of free variables of φ is given by $\mathrm{Free}(\varphi) = \{x_1, \dots, x_n, X_1, \dots, X_m\}$, then the $\mathrm{MSO}(\sigma)$-formula

$$\forall x_1, \dots, x_n \, \forall X_1, \dots, X_m \, \varphi$$

is called the **universal closure** of φ and is denoted by $\forall \varphi$. We note that $\mathrm{Free}(\forall \varphi) = \emptyset$, i.e. $\forall \varphi$ is an $\mathrm{MSO}(\sigma)$-sentence. If φ is itself an $\mathrm{MSO}(\sigma)$-sentence, then we have $\forall \varphi = \varphi$.

A.1.17 Remark. Let φ be an $\mathrm{MSO}(\sigma)$-formula.

a) If $\mathrm{Free}(\varphi)$ contains at least one first-order variable, then $\forall \varphi$ is a first-order universal quantification, and thus we obtain $\emptyset \models \forall \varphi$ (see Remark A.1.13b)).

b) If $\mathfrak{A}$ is a non-empty σ-structure, then one can easily verify that we have $\mathfrak{A} \models \forall \varphi$ if and only if $(\mathfrak{A}, \tau) \models \varphi$ for any finite set $\mathcal{V} \subseteq \mathrm{Var}$ of variables such that $\mathrm{Free}(\varphi) \subseteq \mathcal{V}$ and any $(\mathcal{V}, \mathfrak{A})$-assignment τ (cf. [30, page 40 f.]). Therefore, if $\mathfrak{A} \models \forall \varphi$, then we say that the σ-structure $\mathfrak{A}$ **satisfies** or **models** the formula φ, in symbols $\mathfrak{A} \models \varphi$.

A.1.18 Definition. Let φ and ψ be $\mathrm{MSO}(\sigma)$-formulas. We say that φ and ψ are **(logically) equivalent**, in symbols $\varphi \equiv \psi$, if they fulfill

$$(\mathfrak{A}, \tau) \models \varphi \; \Leftrightarrow \; (\mathfrak{A}, \tau) \models \psi$$

for any σ-structure $\mathfrak{A}$ and any $(\mathcal{V}, \mathfrak{A})$-assignment τ, where $\mathcal{V} = \mathrm{Free}(\varphi) \cup \mathrm{Free}(\psi)$.

A.1.19 Notation.

a) Since the logical operator $\vee$ is associative, we abbreviate $(\dots((\varphi_1 \vee \varphi_2) \vee \varphi_3) \vee \dots \varphi_n)$ by $(\varphi_1 \vee \varphi_2 \vee \dots \vee \varphi_n)$. Likewise, we treat $\wedge$.

b) Given $n \in \mathbb{N}$ and MSO(σ)–formulas $\varphi_1, \dots, \varphi_n$, we use the following standard abbreviations for finite disjunctions and conjunctions:

$$\bigvee_{i=1}^{n} \varphi_i := \begin{cases} \bot & \text{if } n = 0 \\ \bigvee_{i=1}^{n-1} \varphi_i \vee \varphi_n & \text{otherwise,} \end{cases}$$

$$\bigwedge_{i=1}^{n} \varphi_i := \begin{cases} \top & \text{if } n = 0 \\ \bigwedge_{i=1}^{n-1} \varphi_i \wedge \varphi_n & \text{otherwise.} \end{cases}$$

c) Given a finite index set $I = \{i_1, \dots, i_n\}$ with $n \in \mathbb{N}$ and an MSO(σ)–formula φ_i for any $i \in I$, we set:

$$\bigvee_{i \in I} \varphi_i := \bigvee_{k=1}^{n} \varphi_{i_k},$$

$$\bigwedge_{i \in I} \varphi_i := \bigwedge_{k=1}^{n} \varphi_{i_k}.$$

As the logical operators $\vee$ and $\wedge$ are both commutative, the semantics of the formulas $\bigvee_{i \in I} \varphi_i$ and $\bigwedge_{i \in I} \varphi_i$ are independent of the enumeration of the index set I.

d) One can easily verify that $\bot$ is neutral with respect to disjunction. More precisely, we have

$$\varphi \vee \bot \equiv \varphi \quad \text{and} \quad \bot \vee \varphi \equiv \varphi$$

for any MSO(σ)–formula φ. Analogously, $\top$ is neutral with respect to conjunction. Therefore, we abbreviate the formulas

$$\varphi \vee \bot, \ \bot \vee \varphi \quad \text{as well as} \quad \varphi \wedge \top, \ \top \wedge \varphi$$

just by φ. Furthermore, this neutrality justifies our definition of empty disjunctions and conjunctions in b).

A.1.20 Remark. Including also conjunction and universal quantifications in the syntax of MSO(σ), the negation of any MSO(σ)–formula is equivalent to an MSO(σ)–formula in which negation is only applied to atomic formulas. More precisely, for any MSO(σ)–formula φ there exists an MSO(σ)–formula φ^- (possibly containing conjunctions and universal quantifications) in which negation is only applied to atomic formulas and which fulfills the conditions

$$\text{Free}(\varphi^-) = \text{Free}(\varphi) = \text{Free}(\neg \varphi) \quad \text{as well as} \quad \varphi^- \equiv \neg \varphi.$$

Indeed, φ^- can be constructed by structural induction exploiting the following equivalences:

$$\neg\neg\psi \equiv \psi$$
$$\neg(\psi_1 \vee \psi_2) \equiv \neg\psi_1 \wedge \neg\psi_2$$
$$\neg(\psi_1 \wedge \psi_2) \equiv \neg\psi_1 \vee \neg\psi_2$$
$$\neg\exists x\,\psi \equiv \forall x\,\neg\psi$$
$$\neg\forall x\,\psi \equiv \exists x\,\neg\psi$$
$$\neg\exists X\,\psi \equiv \forall X\,\neg\psi$$
$$\neg\forall X\,\psi \equiv \exists X\,\neg\psi$$

For instance, if we consider the formula φ which is given by

$$\forall X \exists x (R(x,y) \wedge \neg x \in X),$$

then its negation $\neg\varphi$ is equivalent to the formula φ^-, which is given by

$$\exists X \forall x (\neg R(x,y) \vee x \in X).$$

Consequently, given an arbitrary formula φ, there exists an equivalent formula ψ in which negation is only applied to atomic formulas and which fulfills $\mathrm{Free}(\psi) = \mathrm{Free}(\varphi)$. Indeed, we obtain ψ by successively replacing every negation of a non-atomic subformula ϕ of φ by the equivalent formula ϕ^- (from front to back).

A.2. Monadic Second-Order Logic for Words

We now apply the notions from Appendix A.1 to a specific signature, which enables us to make logical statements about words. Thus, this appendix complements Section 2.4. We mainly follow Droste [11, § 4]. However, we also took inspiration from Libkin [22, § 7.4] and Thomas [42, § 2 and § 3].

Let Σ be an alphabet.

A.2.1 Definition. We consider the index set $I = \{\Sigma\} \cup \Sigma$ and the purely relational signature $\sigma_\Sigma := (R_i)_{i \in I}$, where the relation symbol

$$R_i \text{ is written } \begin{cases} \leq & \text{if } i = \Sigma \\ P_a & \text{if } i = a \in \Sigma \end{cases}$$

for $i \in I$, and the function $\mathrm{ar}\colon I \to \mathbb{N}^+$ is given by

$$\mathrm{ar}(\leq) = 2,$$
$$\mathrm{ar}(P_a) = 1 \quad (a \in \Sigma).$$

For convenience, we write $\sigma_\Sigma = (\leq, (P_a)_{a \in \Sigma})$ instead of $\sigma_\Sigma = (R_i)_{i \in I}$.

A.2.2 Definition. We simply write $\mathrm{FO}(\Sigma)$ and $\mathrm{MSO}(\Sigma)$ for $\mathrm{FO}(\sigma_\Sigma)$ and $\mathrm{MSO}(\sigma_\Sigma)$, respectively. Moreover, we refer to $\mathrm{FO}(\sigma_\Sigma)$–formulas and $\mathrm{MSO}(\sigma_\Sigma)$–formulas as $\mathrm{FO}(\Sigma)$–formulas and $\mathrm{MSO}(\Sigma)$–formulas, respectively. Thus, the atomic $\mathrm{MSO}(\Sigma)$–formulas are given by:

- $x \leq y$ for any $x, y \in \mathrm{Var}_1$,

- $P_a(x)$ for each $a \in \Sigma$ and $x \in \mathrm{Var}_1$,

- $x \in X$ for any $x \in \mathrm{Var}_1$ and $X \in \mathrm{Var}_2$.

In contrast to Definition 2.4.3, following Definition A.1.3 and Definition A.2.2, the constant $\bot$ is included into thy syntax of $\mathrm{MSO}(\Sigma)$. However, this is not an essential difference due to the fact that $\bot$ can be expressed by an $\mathrm{MSO}(\Sigma)$–formula in the sense of Definition 2.4.3 (see Remark A.2.6).

A.2.3 Notation. Given two individual variables x, y, we usually write

- $x = y$ for the $\mathrm{FO}(\Sigma)$–formula $(x \leq y \wedge y \leq x)$,

- $x < y$ for the $\mathrm{FO}(\Sigma)$–formula $(x \leq y \wedge \neg y \leq x)$.

We usually treat $=$ and $<$ as binary relation symbols, i.e. for instance, we write $x \neq y$ for $\neg x = y$ as well as $x > y$ for $y < x$ (see Notation A.1.4).

Next, we specify the σ_Σ–structures we consider.

A.2.4 Definition. Let w be a finite word over Σ. Just as in Definition 2.4.4, we regard w as σ_Σ–structure by identifying it with the σ_Σ–structure

$$\underline{w} := (\mathrm{dom}(w), \leq, (P_a^w)_{a \in \Sigma}),$$

where

- the domain of $\underline{w}$ is given by the set $\mathrm{dom}(w) := \{1, \ldots, |w|\}$ of positions in w,

- the interpretation $\leq$ of the binary relation symbol $\leq$ is the usual linear ordering on $\mathrm{dom}(w)$ induced by the one on the natural numbers, i.e. $1 \leq 2 \leq \cdots \leq |w|$,

- for each $a \in \Sigma$, the interpretation P_a^w of the unary predicate symbol P_a is defined as the set
$$P_a^w := \{i \in \mathrm{dom}(w) \mid w(i) = a\}.$$
As usual, we simply write P_a for P_a^w if no confusion is likely to arise.

In particular, the empty word corresponds to the empty σ_Σ–structure $\underline{\varepsilon} = (\emptyset, \emptyset, (\emptyset)_{a \in \Sigma})$.

Classical Model Theory mainly deals with infinite structures, as its origins are in Mathematics and most mathematical objects of interest are infinite (cf. Libkin [22, page 1]). However, we are interested in the expressive power of monadic second-order logic for words, which are finite objects. Therefore, our treatment concentrates on *finite* structures, i.e. structures having a finite universe, as we consider only σ_Σ–structures of the form $\underline{w}$, where w is a word over Σ.

Given a word w over Σ and a finite set $\mathcal{V} \subseteq \mathrm{Var}$ of variables, we refer to $(\mathcal{V}, \underline{w})$–assignments (see Definition A.1.10) as $(\mathcal{V}, w)$–assignments (see Definition 2.4.8). Similarly, we write $(w, \sigma) \models \varphi$ (see Definition 2.4.9) for $(\underline{w}, \sigma) \models \varphi$ (see Definition A.1.12).

A.2.5 Remark. We immediately obtain Lemma 2.4.10 by applying Lemma A.1.15 to the signature σ_Σ.

In Definition 2.4.11 and Definition 2.5.6 we explain how MSO(Σ)–formulas define languages over suitable alphabets.

A.2.6 Remark. We usually omit the "constant" formula $\bot$ in the syntax of MSO(Σ) (see Definition 2.4.3), since over this specific signature, $\bot$ can be replaced by the MSO(Σ)–sentence $\exists x\, x < x$. Indeed, one can easily verify that we have

$$L_\mathcal{V}(\bot) = L_\mathcal{V}(\exists x\, x < x) = \emptyset$$

for any finite set $\mathcal{V} \subseteq \mathrm{Var}$ of variables, where $L_\mathcal{V}(\bot) = \{(w, \sigma) \in N_\mathcal{V} \mid (w, \sigma) \models \bot\}$. In particular, we obtain

$$(w, \sigma) \models \bot \;\Leftrightarrow\; (w, \sigma) \models \exists x\, x < x$$

for any word w over Σ and any $(\mathcal{V}, w)$–assignment σ, where $\mathcal{V} \subseteq \mathrm{Var}$ is a finite set of variables. Similarly, the "constant" formula $\top$ can be expressed by the MSO(Σ)–sentence $\forall x\, x \leq x$. More precisely, we have

$$L_\mathcal{V}(\top) = L_\mathcal{V}(\neg\bot) = L_\mathcal{V}(\neg\exists x\, x < x) = L_\mathcal{V}(\forall x\, x \leq x) = N_\mathcal{V}$$

for any finite set $\mathcal{V} \subseteq \mathrm{Var}$ of variables, where $L_\mathcal{V}(\top) = \{(w, \sigma) \in N_\mathcal{V} \mid (w, \sigma) \models \top\}$. In particular, the characteristic series of the languages $L(\bot) := L_\emptyset(\bot)$ and $L(\top) := L_\emptyset(\top)$ over Σ are constant series with values 0 and 1, respectively, i.e.

$$\mathbb{1}_{L(\bot)} = \sum_{w \in \Sigma^*} 0w \quad \text{and} \quad \mathbb{1}_{L(\top)} = \sum_{w \in \Sigma^*} 1w.$$

A.2.7 Remark. The semantics of the FO(Σ)–formula $(x \leq y \wedge y \leq x)$, which we abbreviate by $x = y$, corresponds to the usual meaning of the identity relation. More precisely, a pair (w, σ) consisting of a word and a $(\mathcal{V}, w)$–assignment, where $\mathcal{V} \subseteq \mathrm{Var}$ is a finite set of variables containing x and y, satisfies the formula $x = y$ if and only if the positions $\sigma(x)$ and $\sigma(y)$ do coincide. Thus, the formula $(x \leq y \wedge y \leq x)$ expresses the equality of two positions. Similarly, the FO(Σ)–formula $x < y$ expresses the usual meaning of the relation $<$.

A.2.8 Remark. Let φ and ψ be MSO(Σ)–formulas. If φ and ψ are equivalent in the sense of Definition A.1.18, then φ and ψ are equivalent in the sense of defining the same language. More precisely, by exploiting Lemma A.1.15 one can verify that

$$\varphi \equiv \psi \;\Rightarrow\; L_\mathcal{V}(\varphi) = L_\mathcal{V}(\psi),$$

where $\mathcal{V} \subseteq \mathrm{Var}$ is a finite set of variables containing $\mathrm{Free}(\varphi) \cup \mathrm{Free}(\psi)$.

The converse of the above implication does in general not hold. For instance, the MSO(Σ)–sentence

$$\varphi := \exists x \ x < x$$

and the MSO(Σ)–sentence

$$\psi := \exists x \ x = x \wedge \forall y \, \exists z \ y < z$$

are equivalent in the sense of defining the same language. More precisely, we have

$$L_\mathcal{V}(\varphi) = L_\mathcal{V}(\psi) = \emptyset$$

for any finite set $\mathcal{V} \subseteq$ Var of variables. On the other hand, for the σ_Σ–structure $\mathfrak{A} = (\mathbb{N}, \leq, (\emptyset)_{a \in \Sigma})$, where $\leq$ is the usual linear ordering on $\mathbb{N}$, we obtain

$$\mathfrak{A} \not\models \varphi \ \text{ and } \ \mathfrak{A} \models \psi.$$

Therefore, the MSO(Σ)–sentences φ and ψ are not equivalent in the sense of Definition A.1.18, i.e.

$$\varphi \not\equiv \psi.$$

In this example it is essential that the universe of the σ_Σ–structure $\mathfrak{A}$ is an infinite set.

References

[1] C. BAIER, M. GRÖSSER and F. CIESINSKI, 'Model Checking Linear-Time Properties of Probabilistic Systems', *Handbook of Weighted Automata* (Eds. M. Droste, W. Kuich and H. Vogler), Monogr. Theoret. Comput. Sci. EATCS Ser. (Springer, Berlin, 2009) 519–570.

[2] J. BERSTEL and C. REUTENAUER, *Rational Series and Their Languages*, EATCS Monogr. Theoret. Comput. Sci. 12 (Springer, Berlin, 1988).

[3] J. R. BÜCHI, C. C. ELGOT and J. B. WRIGHT, 'The nonexistence of certain algorithms for finite automata theory' (Preliminary report), *Notices Amer. Math. Soc.* **5** (1958) 98.

[4] J. R. BÜCHI, 'Weak second-order arithmetic and finite automata', *Z. Math. Logik Grundlagen Math.* **6** (1960) 66–92.

[5] M. DROSTE and P. GASTIN, 'Weighted Automata and Weighted Logics', *Automata, Languages and Programming* (Eds. L. Caires, G. F. Italiano, L. Monteiro, C. Palamidessi and M. Yung), Lect. Notes Comput. Sci. 3580 (Springer, Berlin, 2005) 513–525.

[6] M. DROSTE and P. GASTIN, 'Weighted automata and weighted logics', *Theoret. Comput. Sci.* **380** (2007) 69–86.

[7] M. DROSTE and P. GASTIN, 'On Aperiodic and Star-Free Formal Power Series in Partially Commuting Variables', *Theory Comput. Syst.* **42** (2008) 608–631.

[8] M. DROSTE, W. KUICH and H. VOGLER (Eds.), *Handbook of Weighted Automata*, Monogr. Theoret. Comput. Sci. EATCS Ser. (Springer, Berlin, 2009).

[9] M. DROSTE and W. KUICH, 'Semirings and Formal Power Series', *Handbook of Weighted Automata* (Eds. M. Droste, W. Kuich and H. Vogler), Monogr. Theoret. Comput. Sci. EATCS Ser. (Springer, Berlin, 2009) 3–28.

[10] M. DROSTE and P. GASTIN, 'Weighted Automata and Weighted Logics', *Handbook of Weighted Automata* (Eds. M. Droste, W. Kuich and H. Vogler), Monogr. Theoret. Comput. Sci. EATCS Ser. (Springer, Berlin, 2009) 175–211.

[11] M. DROSTE, 'Automatentheorie' (Lecture Notes, English), University of Leipzig, winter term 2020/2021.

[12] M. DROSTE and D. KUSKE, 'Weighted Automata', *Theoretical Foundations*, Handbook of Automata Theory, Vol. I (Ed. J.-E. Pin, EMS Press, Berlin, 2021) 113–150.

[13] H.-D. EBBINGHAUS, J. FLUM and W. THOMAS, *Mathematical Logic*, 3rd edn., Undergraduate Texts Math. 291 (Springer, Cham, 2021).

[14] S. EILENBERG, *Automata, languages, and machines*, Vol. A, Pure Appl. Math. 58 (Academic Press, New York, 1974).

[15] C. C. ELGOT, 'Decision problems of finite automata design and related arithmetics', *Trans. Amer. Math. Soc.* **98** (1961) 21–51.

[16] J. E. HOPCROFT, R. MOTWANI and J. D. ULLMAN, *Introduction to Automata Theory, Languages, and Computation*, 3rd edn. (Pearson Education, Boston, MA, 2007).

[17] B. KHOUSSAINOV and A. NERODE, *Automata Theory and its Applications*, Prog. Comput. Sci. Appl. Log. 21 (Springer, New York, 2001).

[18] S. C. KLEENE, 'Representation of events in nerve nets and finite automata', *Automata studies* (Eds. C. E. Shannon and J. McCarthy), Ann. Math. Stud. 34 (Princeton University Press, Princeton, NJ, 1956) 3–41.

[19] L. S. KRAPP, S. KUHLMANN and M. SERRA, 'On Rayner structures', *Commun. Algebra* **50** (2022) 940–948.

[20] S. KUHLMANN, 'Real Algebraic Geometry II' (Lecture Notes), University of Konstanz, summer term 2019.

[21] W. KUICH and A. SALOMAA, *Semirings, Automata, Languages*, EATCS Monogr. Theoret. Comput. Sci. 5 (Springer, Berlin, 1986).

[22] L. LIBKIN, *Elements of Finite Model Theory*, Texts Theor. Comput. Sci. EATCS Ser. (Springer, Berlin, 2004).

[23] B. H. NEUMANN, 'On ordered division rings', *Trans. Amer. Math. Soc.* **66** (1949) 202–252.

[24] W. S. MCCULLOCH and W. PITTS, 'A logical calculus of the ideas immanent in nervous activity', *Bull. Math. Biophys.* **5** (1943) 115–133.

[25] R. MCNAUGHTON and S. PAPERT, *Counter-free automata*, MIT Research Monographs 65 (MIT Press, Cambridge, MA, 1971).

[26] D. PERRIN and J.-E. PIN, 'First-Order Logic and Star-Free Sets', *J. Comput. Syst. Sci.* **32** (1986) 393–406.

[27] D. PERRIN, 'Finite Automata', *Formal Models and Semantics*, Handbook of Theoretical Computer Science, Vol. B (Ed. J. van Leeuwen; Elsevier, Amsterdam; MIT Press, Cambridge, MA, 1990) 1–57.

[28] J.-E. PIN (Ed.), *Theoretical Foundations*, Handbook of Automata Theory, Vol. I (EMS Press, Berlin, 2021).

[29] J.-E. PIN (Ed.), *Automata in Mathematics and Selected Applications*, Handbook of Automata Theory, Vol. II (EMS Press, Berlin, 2021).

[30] A. PRESTEL and C. N. DELZELL, *Mathematical Logic and Model Theory: A Brief Introduction*, Universitext (Springer, London, 2011).

[31] M. O. RABIN and D. SCOTT, 'Finite Automata and Their Decision Problems', *IBM J. Res. Develop.* **3** (1959) 114–125.

[32] G. RAHONIS, 'Fuzzy Languages', *Handbook of Weighted Automata* (Eds. M. Droste, W. Kuich and H. Vogler), Monogr. Theoret. Comput. Sci. EATCS Ser. (Springer, Berlin, 2009) 481–517.

[33] J. SAKAROVITCH, *Elements of automata theory* (Cambridge University Press, Cambridge, 2009).

[34] J. SAKAROVITCH, 'Rational and Recognisable Power Series', *Handbook of Weighted Automata* (Eds. M. Droste, W. Kuich and H. Vogler), Monogr. Theoret. Comput. Sci. EATCS Ser. (Springer, Berlin, 2009) 105–174.

[35] A. SALOMAA and M. SOITTOLA, *Automata-Theoretic Aspects of Formal Power Series*, Texts and Monographs in Computer Science (Springer, New York, 1978).

[36] A. SALOMAA, 'Formal Languages and Power Series', *Formal Models and Semantics*, Handbook of Theoretical Computer Science, Vol. B (Ed. J. van Leeuwen; Elsevier, Amsterdam; MIT Press, Cambridge, MA, 1990) 103–132.

[37] M. P. SCHÜTZENBERGER, 'On the Definition of a Family of Automata', *Inf. Control* **4** (1961) 245–270.

[38] M. P. SCHÜTZENBERGER, 'On a Theorem of R. Jungen', *Proc. Amer. Math. Soc.* **13** (1962) 885–890.

[39] M. P. SCHÜTZENBERGER, 'On Finite Monoids Having Only Trivial Subgroups', *Inf. Control* **8** (1965) 190–194.

[40] E. D. SONTAG, 'On Some Questions of Rationality and Decidability', *J. Comput. Syst. Sci.* **11** (1975) 375–381.

[41] H. STRAUBING, *Finite Automata, Formal Logic, and Circuit Complexity*, Progress in Theoretical Computer Science (Birkhäuser, Boston, MA, 1994).

[42] W. THOMAS, 'Languages, Automata, and Logic', *Beyond Words*, Handbook of Formal Languages, Vol. 3 (Springer, Berlin, 1997) 389–455.

[43] B. A. TRAKHTENBROT, 'Finite automata and the logic of single-place predicates', *Dokl. Akad. Nauk SSSR* **140** (1961) 326–329 (Russian), *Sov. Phys. Dokl.* **6** (1962) 753–755 (English).

Index

© The Editor(s) (if applicable) and The Author(s), under exclusive license to
Springer Fachmedien Wiesbaden GmbH, part of Springer Nature 2022
L. Wirth, *Weighted Automata, Formal Power Series and Weighted Logic*,
BestMasters, https://doi.org/10.1007/978-3-658-39323-6

Printed in the United States
by Baker & Taylor Publisher Services